HOW GLASSES CAUGHT A KILLER

And other stories of how optics changed the world

David Baker

David Baker.

Published in 2016 by FeedARead.com Publishing

Second Edition

A CIP catalogue record for this title is available from the British Library.

CONTENTS

4. Sport and Miscellany

PREFACE

This volume grew out of *Optical Connections* which began life as a short series of articles that looked at a few topics from the point of view of their tangential, and sometimes more direct, links with the world of optics. The idea was to provide *Optician* magazine with something a little different from the clinical articles and optical industry news that is the beating heart of the journal; something to take the mind off work for a moment, to relax with over a cup of tea, perhaps. As the series went on, new subject matter kept presenting itself and, fortunately, *Optician*'s response was, 'Keep it coming!' for which I am grateful.

Material for articles appeared from the most unexpected places. Sometimes, people suggested ideas to me: my wife, my parents, Seth Belson and Frank Norville (apologies if I have forgotten anyone). Often it has been a case of rooting around in libraries and finding an interesting morsel in a neglected volume. A lecture on Michael Faraday introduced me to his optical work which had been unknown to me, and a chance meeting afterwards with the speaker, Professor Frank James, in the Gents of all places, led to him kindly sending me material for *The Supreme Experimenter*. During a visit to the National Railway Museum, York, for the benefit of my train-mad sons, my wife noticed a curious snippet of information that resulted in *The Engineer's New Coat*.

Coincidences arising from some articles are stories in themselves. How was I to know, for instance, when researching *The Engineer's New Coat*, that a friend's wife would just happen to be a descendant of the engineer in question, William Stroudley? Or that shortly after publication of *Voyage Into Darkness* an episode of BBC's *Garrow's Law* would feature part of its story. Or that, when researching John Cuff for *Old Master in Optics*, Zoffany's painting of him was soon to have a rare public showing at the Royal Academy?

Some of the pieces are more technical than others, hopefully, there is enough technical content for those with an optical or general scientific interest, but not so much as to put off those not so inclined.

Inevitably there are people I need to thank for their help. Mostly they are acknowledged at the end of specific articles they have assisted with. But I would especially like to highlight the long-term encouragement of Frank Norville, chairman of Norville Optical Co and the unfailing assistance of Mike Hale, *Optician* Features and Education Editor. Above all, though, I thank my wife, Julie, for her constant patience and support.

David Baker
November 2015

1. War and Conflict

HOW GLASSES CAUGHT A KILLER

The perfect crime – almost; and the 'trial of the century'

This is the story of what is commonly referred to as the twentieth century's first 'trial of the century'; a story of two cold-hearted killers, one of the most famous defence attorneys in American legal history and the chance discovery of a simple pair of spectacles that led to the murderers' conviction.

The setting is Chicago, 1924, in the affluent neighbourhood of South Greenwood, only a block away from where Barack Obama's residence now sits, on the favoured south side of the city. In that suburb lived two wealthy families. Each had a teenage son as brilliant intellectually as their personalities were troubled. For all their anti-social tendencies, it is generally accepted that it was only as a result of their meeting and forming of a complex relationship that such a heinous crime of premeditated murder could have been committed.

Richard Loeb was the second of four sons of a vice-president of the Sears Roebuck mail-order corporation. His governess recognised his early academic promise, pushing him hard in his studies. He came to resent the enforced workload; nevertheless he graduated from high school at fourteen and, in 1923, became the University of Michigan's youngest ever graduate, aged seventeen. He had been popular with his peers, but his fascination with detective and crime novels led him to start committing crimes, mostly criminal damage, for thrills.

Nathan Leopold, Jr could not have been more different in character than the extrovert, seemingly easygoing Loeb. Leopold was self-conscious about his somewhat awkward features, but his prodigious intelligence led many to see him as outwardly contemptuous and arrogant. He was completely spoiled by his industrialist father after his mother died when Nathan was young. He devoured knowledge, becoming fluent in at least nine languages, a noted botanist, a national authority on ornithology and a classics and philosophy scholar – all by the age of eighteen. It was the latter, in particular his fascination with Nietzsche's idea of the 'superman', the superior man who rises above the conventional morality, that was to cause his downfall.

Leopold was preparing to enter the University of Chicago when he met Loeb, already a student there prior to his transfer to Michigan. They were an unlikely pair, but Leopold fell completely under the spell of the charming, handsome Loeb. Despite having a girlfriend at the time, Leopold was sexually attracted to Loeb, and accepted the role of Loeb's partner in petty crime in return for his sexual favours. Despite drifting apart for a time, they renewed their friendship when both re-enrolled at Chicago, Loeb on a history course and Leopold in the law school. Their activities soon escalated to theft and arson, Loeb for the thrill of committing undetected crimes, Leopold for the sex and the desire to live the Nietzschean life above morals.

The boys' relationship proved to be the spur for an idea Loeb began to develop for committing the perfect crime. The plan they came up with was for the kidnapping of a son of wealthy parents and demand for a large ransom. They soon realised that, though unpalatable, to be sure of evading detection the boy would have also to be killed. Leopold wrote later that his motive for acquiescing to the scheme 'to the extent that I had one, was to please Dick.' He explained that, 'Loeb's friendship was necessary to me – terribly necessary.'

On the afternoon of May 21, 1924, the pair were cruising in a rented car looking for a victim. They spotted fourteen year-old Bobby Franks, whom Loeb was acquainted with, walking home from school. Loeb enticed him into the car on a pretext, and when they drove off he was killed by several blows to the head with a chisel; most evidence suggests the murderer was Loeb. They drove to a marshland that Leopold frequented for birdwatching, stripped the body naked, poured hydrochloric acid over it and dumped it in a concrete drainage culvert. The clothes were taken away and burned. That evening Leopold phoned Mrs Franks to tell her that Bobby had been kidnapped, unharmed, and to expect a ransom note. A letter duly arrived the next day instructing the Franks' to gather $10,000 in old, unmarked bills and to await further instructions. Mid-afternoon, Mr Franks received another call telling him to take the money in a taxi shortly to arrive, to a named drugstore. But before he could comply another call came, from the police, to say that Bobby's body had been discovered, a chance sighting by a labourer.

The boys may never have been caught but for 'the hand of God at work in this case' according to the prosecutor. A pair of tortoiseshell spectacles were found at the scene which, at first, seemed quite ordinary. When Leopold read of the find, he became worried when he could not locate his similar-looking pair in the jacket pocket in which he kept them. He had only worn them for a few weeks months earlier to cope with headaches. He consoled himself, as he recounted later, telling Loeb, 'I know the prescription is a very common one. The doc told me so. And how are they going to know what oculist they come from? They'd have to go through the records of every oculist in town and then check on a couple of thousand people.'

But on closer inspection the frame had an unusual hinge which enabled it to be traced to a single Chicago optometrist, Emil Deutsch. The hinge was manufactured by a New York company that had only one Chicago outlet, Almer Coe & Co, and they had sold only three of the frames with this hinge. One wearer was female, another travelling in Europe at the time, the third belonging to Leopold. The police brought him in to be interviewed by State's Attorney Robert Crowe, whereat he claimed that the spectacles must have fallen out of his pocket when he tripped on one of his regular birding expeditions. But after several attempts at a demonstration he failed to dislodge them.

Two other pieces of evidence then sealed the boys' fate. Firstly, a typewriter that Leopold used at college was found to match the type on the ransom note. Secondly, the boys claimed that on the evening of the murder they had been driving around with some girls in the Leopold family car; ironically, in a misguided effort to clear Leopold, the family chauffeur came forward to state that the car had not left the garage that day.

The families hired Clarence Darrow to defend the boys. If this was the case that really made his name, he became even more famous a year later as the defence lawyer in the 'Scopes Monkey Trial'. The prosecution's plan was to try the boys first for murder and then, if found innocent, for kidnapping; both crimes attracting the death penalty. Darrow got them to plead guilty so that the state would get only one go at the death sentence, and at a hearing by the judge rather than a jury. A twelve-hour summation by Darrow followed, largely a moral crusade against the death penalty. He 'won' a sentence of life (for murder) plus 99 years (kidnapping).

Loeb would be attacked and killed in jail in 1936. Leopold was released after serving 34 years. To escape publicity he emigrated to Puerto Rico, where he eventually married despite still declaring his love for Loeb. Leopold died in 1971, following which his wish to donate his corneas was carried out.

A FAMILY OF HEROES

The incredible gallantry of an ophthalmologist and his brothers

For those within optics, the word 'Chavasse' is mostly synonymous with a frosted lens (although a Chavasse lens is of superior cosmetic appearance). But what a story lies behind that little-considered item! In fact this occluder bears the name of an ophthalmic surgeon who also edited a classic text book, devised new surgical instruments and committed at least one act of outstanding bravery during the First World War. Yet even his wartime exploits were outshone by one of the brothers in his quite extraordinary family.

Francis Bernard Chavasse (known as Bernard) was born on December 2, 1889, the second youngest of a family of seven children that contained two sets of twins. The family name derived from a French ancestor who had come to England and become involved in the 1715 Jacobite Rebellion. Although a Roman Catholic, within two generations the family had converted to Protestantism. The Church figured prominently in the Chavasse family, with Bernard's father becoming Bishop of Liverpool and instilling a strong Christian faith in his offspring. Surgeons featured prominently too, while other members went into the Army and the Law. Medicine and the military were to feature in Bernard's life, as also they were to combine heroically - and tragically - for his older brother, Noel.

Bernard followed Noel and the latter's twin brother, Christopher, to Oxford, but studying at Balliol rather than his brothers' college, Trinity. Together with Noel he joined the university's Officer Training Corps. After graduating with a First in Natural Sciences - as did Noel - in 1912 (winning the prize in Anatomy along the way), he continued his medical studies at Liverpool University in the city in which he grew up. On the outbreak of war he volunteered to join the Royal Army Medical Corps (RAMC) and was then transferred to the First King's Liverpool Regiment.

There he served as a medical officer in Egypt and Gallipoli, before arriving on the Western Front in 1916. By now all four male Chavasse siblings were serving there: Noel's twin, Christopher, as an army chaplain, Noel and Bernard as medical officers and the youngest child, Aidan, who served in Bernard's battalion.

All the boys were athletic and had a strong sense of duty. The twins, Noel and Christopher, represented Oxford University at athletics and lacrosse and ran in the 1908 Olympic Games, while Aidan won sports honours at his school. All were conspicuous in their bravery; yet they saw their acts as nothing more than their duty to help their fellow soldiers. Noel gained a Military Cross (MC) in May 1915, followed by the award of the Victoria Cross for his actions in rescuing wounded soldiers while under heavy fire at Guillemont in August 1916. Then Christopher and Bernard were both awarded an MC, although Bernard revealed in a letter to his sister, Marjorie, that his superiors had actually nominated him for a VC. Some felt Christopher's actions also merited a VC; but he was honoured with the Croix-de-Guerre for his distinguished service.

Both Noel and Bernard carried out incredible feats of valour time and again to rescue and treat the wounded under conditions of extreme danger with no thought for their own wellbeing. After yet another selfless act, Noel was seriously injured and later died from his wounds. In the midst of his grief at the news, his father received word that Noel was to receive posthumously a Bar to his

VC, thus becoming the only man to be honoured with two VCs in that war. One wonders how the Chevasse family coped with their loss, given that not long before, Aidan had been reported wounded and missing in action. The official report stated that Lt Chavasse had led eight men on a patrol when they were fired upon; Aidan was injured and sent his men back without him. Bernard went out twice to try to locate him without success. He was never found. Within a period of two months, the Chevasse family had gained two MCs, a second VC but, sadly, lost two sons.

After the war, Christopher helped his father to establish St Peter's Hall, the aim being to provide an opportunity for students of limited means to gain an Oxford education. He later progressed in the Church of England to become Bishop of Rochester, and became a leading light of its Evangelical wing. He was able to witness St Peter's become a full college of Oxford University just a year before his death in 1962.

Bernard returned to Liverpool, where he specialised in ophthalmology, in which he excelled. The papers he delivered to the Ophthalmological Society and at the Northern and Oxford Congresses cemented his reputation. That reputation led to an invitation to edit a new edition of the classic text, *Worth's Squint* when the author's health began to fail. Applying his ideas relating to new developments in physiology and pathology he largely re-wrote the book, bringing it up to date and including descriptions of some of his own surgical procedures. He challenged Worth's prevailing theory of an innate weakness of a faculty for fusion being responsible for the inability of some children to attain binocular vision. Instead he proposed that the inability to fuse was the consequence of abnormal development of binocular reflexes, and that surgery to correct a squint and align the eyes could allow proper development of these reflexes to occur. Chavasse was not a great believer in orthoptics as treatment, but he did feel that, applied rigorously and systematically, orthoptic investigation could be a useful aid in diagnosis and for binocular re-education between stages of surgery. The seventh edition of the book, published in 1939, became *Worth and Chavasse's Squint*.

In addition to novel surgical techniques, Chavasse invented several surgical instruments, including fixation forceps, a strabismus hook and a marginal myotomy retractor. His best-known device, though, is his eponymous lens. It is often used as an alternative term for a frosted lens, but the 'hammered' surface of a Chavasse lens is less opaque. The British Standards definition of a Chavasse lens is as follows: 'A lens with an irregular surface used to depress the visual acuity while permitting the eye to be seen from the front'.

A few words deserves to be said about the Chavasse sisters. Dorothea, the eldest of the seven children, died in 1935, a rector's wife, aged 52. Edith Marjorie (known as Marjorie) and Mary Laeta (May) were a second set of twins, born after Noel and Christopher. They both volunteered for nursing duties during WWI at a convalescent home run by their aunt. Marjorie was a source of supplies to her brothers in France and dealt with some financial matters on behalf of Noel's men. She later worked for Barnardo's. May travelled to France to assist at a mobile field hospital, receiving a mention in dispatches. She later qualified as a nurse and was active during the Second World War as a member of the Queen Alexandra's Military Nursing Service. Marjorie and May celebrated their 100th birthdays in 1986, passing away one and three years later respectively.

Keith Lyle, who took over editorship of *Worth and Chavasse's Squint* for the eighth edition, says in his Preface that 'Chavasse has unquestionably made a valuable and original contribution to ophthalmology and has placed the aetiology, diagnosis and treatment of strabismus on a sound physiological basis. Ophthalmology owes Bernard Chavasse a great debt of gratitude.' The reason Lyle had taken over? Sadly, Chavasse was killed in a car accident in 1941, aged 52.

A FATAL CONNECTION

The optician and the St Valentine's Day massacre

They say career advancement is all about networking; meeting the right people and making the right choices. And the people with whom we associate will partly define our professional standing. But one American optician made some very bad career choices indeed. So bad, that the consequences were to prove fatal to him one morning in 1929, the 14th of February.

In his business advertisements Reinhart Schwimmer styled himself 'Dr R Schwimmer, optometrist and eyesight specialist' even though he apparently had little or no medical training. His practice was at 625 North Avenue, Chicago, opening hours supposedly 10am – 8pm. In fact he was an inveterate gambler, running up large debts; and he spent more time at the racetrack than at his work, causing his practice to founder. His gambling also brought him into contact with the more unsavoury types in Chicago society, and so began his fascination with that city's Prohibition-era gangsters.

Schwimmer was never a gangster himself but liked to hang out with members of George 'Bugs' Moran's North Side Irish gang for the thrill of being associated with those characters. He liked to boast to his friends that he was in the illicit booze business, spinning them tall tales of importing alcohol from Detroit, and would claim that, with his connections, he could have anyone 'whacked' should he wish it. These fantasies, and his gambling, also contributed to a convoluted private life.

Reinhart Schwimmer's birthdate is usually given as 1 December 1900 (the date shown on his headstone), although his birth certificate says 1899. His first name is normally spelt 'Reinhardt' but is engraved without the 'd' on the headstone. He was certainly born in Chicago and married Fae Johnson, whom he divorced in 1923. He then moved into the Parkway Apartments but, when his debts accumulated, he was asked to leave. Having found a rich widow, Mrs Risch, to marry, he was able to move back into the Parkway on the understanding that his gangland associations were finished. When she divorced him in 1927-8, he went back to spending his days with his North Side friends.

Schwimmer might have stayed on the periphery of gangland violence were it not for his penchant for emulating the dress style and appearance of his hero, 'Bugs' Moran. One of Moran's main rivals for business was the South Side Italian gang, led by a certain Al Capone. Capone was mad at Moran for the North Side's frequent hijacking of his Canadian alcohol en route to Chicago, as well as for the their execution of South Side gang members. On Capone's orders, a plan was devised to lure Moran to the North Side's hang-out, where Jack 'Machine Gun' McGurn would carry out the hit.

The commonly-held theory is that Moran was to expect a delivery of cut-price whiskey to the North Side headquarters, the SMC Cartage warehouse at 2122 North Clark Street at 11am on 14 February. Schwimmer wanted to come along for the ride, and accompanied gangsters Frank and Pete Gusenberg, James Clerk, Adam Heyer and Al Weinshank to the warehouse, where gang mechanic, John May, was working on a car. Moran was to meet them there. At 10.30am, tipped off by lookouts who may have mistaken Schwimmer for Moran, a police car pulled up outside the warehouse. Four men emerged, two in police uniform, and entered the building. The 'policemen' frisked, disarmed and lined Moran's men up against a wall. The gangsters were unconcerned with this as they were well used to a police rousting and assumed that, if arrested, their connections would soon gain them their release. Schwimmer was probably looking forward to having his name in the newspaper reports as vindication of his bona fide gangland connections.

The fake officers now signalled to their plain-clothed accomplices to enter the room with their Thompson sub-machine guns which they proceeded to fire at the seven men, spraying them with 70 bullets. Only Frank Gusenberg, despite 14 bullet wounds, was not killed outright, although he died three hours later still claiming, 'Nobody shot me.' The two 'policemen' made a show of escorting the killers from the building as if under arrest and escaped in the police car. The hit was a failure in that Capone didn't get Moran: some say he was simply late showing up, others that he saw the police car and coolly drove on.

So ended what became known as the St Valentine's Day Massacre. But the unfortunate fantasist and sometime optician, Schwimmer, would surely have approved of the Chicago Police Department's homicide record's description of him as 'One of the seven Moran gangsters...'

My thanks to optometrist Seth Belson for his assistance with this article.

FORGED BY WAR

How the plastic in your spectacle lenses was born

A man born in a log cabin in Danville, Kentucky, had his life shaped, and made his fortune, through war. He went on to found a company that produced a chemical whose first – and at the time, possibly only – use was in war. That chemical has become perhaps the most important and ubiquitous material in the optical industry worldwide.

The man was John Baptiste Ford (1811 – 1903). His mother was the daughter of a French immigrant who fought in the American War of Independence. His father joined the Kentucky Homespun Volunteer Regiment to fight the British at New Orleans in the War of 1812 and never returned, so Ford never properly knew him. Apprenticed to a saddlemaker at the age of twelve, he ran away two years later to Grenville, Indiana. There, in 1831, he married Mary Bower who taught him to read and write. They opened a dry goods store, then a saddler shop and a flour mill.

Ford's first big break was winning a contract to supply the US Cavalry during the American-Mexican War of 1846-8. On the back of that he built a successful boatworks at New Albany, Indiana, earning him the nickname 'Captain'. Next he entered the iron industry, supplying the Union forces during the American Civil War. When one of his sons, Emory, was a student in Pittsburgh, he was impressed by the many glassworks there. Ford opened his own glassworks in New Albany, expanding into plate glass - then a mainly European-manufactured product. In 1883 Ford and sons formed a new company in Pittsburgh: the Pittsburgh Plate Glass Company. By 1900 the company had a 65 per cent share of the US plate glass market, and was the second-largest US producer of paints. It grew into a huge conglomerate that became, in 1965, PPG Industries, to reflect its diverse business interests. A lasting tribute to that humble illiterate made good is the town of Ford City, Pennsylvania, named in his honour.

In the late 1930s, PPG was looking for a way to create an allyl resin with low-pressure thermostatic properties. Plastics such as polystyrene and nylon had been developed but with the onset of World War II the scarcity of natural resources made the search for new materials the more urgent.

Other thermoplastic materials such as Plexiglas and Lucite (both basically PMMA) were already available. PPG used a subsidiary company in Ohio, Columbia Southern Chemical Co to research a new clear resin. By May 1940, their 39th attempt, one particular compound, an allyl diglycol carbonate (ADC) monomer was finally showing promise. It had unique characteristics, including the ability to combine with multiple layers of cloth, paper and other materials to produce exceptionally strong laminated products capable of being moulded into a variety of reinforced shapes: the new industry of 'reinforced plastics' was born. PPG trademarked their new material under its batch name.

The new invention was soon found a role in helping the war effort. The B-17 bomber was a formidable machine, but had two weaknesses. Firstly, it was vulnerable to being shot through its fuel tank. Secondly, transparent inspection tubes embedded in the fuel lines running through the engineers' compartment (providing the crew with a visible fuel gauge) were glass and often shattered during combat, spraying gasoline throughout the cockpit. The new resin combined with fibreglass to produce a much lighter fuel tank, thus extending the bomber's range and, when lined with a self-sealing rubber compound, was safer too. It also replaced the glass in the inspection tubes. The B-17 could now truly deserve its nickname of the Flying Fortress.

At the end of the war all government contracts were cancelled and PPG was left with a steel tank containing 38,000 lb of resin with no obvious use. It was known to remain liquid until a catalyst was added, but would eventually self-harden. No-one knew how quickly this would occur: would PPG end up with a useless steel-encased slab of plastic? A frantic search ensued for potential clients for the remaining batch. Although unknown at the time, it turns out that the resin monomer is remarkably stable, with a very long shelf-life, and the leftover material never solidified.

A sales manager at Univis Lens Co, Robert Graham, had previously heard about the resin even though it had been classified as a military property during the war and, somehow, had obtained five gallons of it to experiment with. Pre-war, the Unbreakable Lens Company of America (TULCA), later acquired by Univis, had been working on developing plastic lenses. They had looked at polymethyl methacrylate (PMMA – best known by the trade name, Perspex), following on from Combined Optical Industries Ltd (COIL) in the UK who had distributed their PMMA lenses in the US as 'Igard'. But these lenses scratched easily and yellowed even when out of direct light.

When Univis closed down TULCA on abandoning their plastic lens research, Graham took most of Univis' research team and set up his own company in California, the Plastic Lens Co, later to become Armolite. Their PMMA lenses were optically clear, but scratching was still a problem. Now Graham negotiated for PPG's tank of resin, deciding to try it out as a possible material for ophthalmic lenses. The results were very promising. It was optically clear but thirty times more abrasion-resistant than Plexiglas or Lucite. There were problems to overcome, however. The property of the resin as a bonding agent for laminated materials that was so useful in wartime was a distinct disadvantage for lens casting as the resin tended to bond with the moulds. This was especially true of metal moulds but, fortunately, Graham's group favoured glass ones. Another difficulty was that the lenses cast suffered 14 percent shrinkage as the material cured. With prescription lenses the variation in thickness between the edge and centre resulted in differential shrinkage, creating optical distortions in the finished lenses. Graham's solution was to cast thick blanks such that the back curve matched the finished front curve. Uniform shrinkage was thus achieved during curing, after which the back surface could be ground and surfaced as required.

Once the moulding and shrinkage problems had been overcome, the newly-incorporated Armolite Co began producing their resin lenses in 1947. It had a worldwide monopoly for six years, after which Essilor in France, followed by SOLA in Australia and then American Optical began producing lenses of the same material. Univis eventually became a major plastic lens producer. Until 1961 PPG's primary resin sales were for flat sheet applications such as welding helmets, industrial goggles, gas masks and windscreens for industrial vehicles. But by 1975 over 90 percent of PPG's resin sales were to the optical industry.

Despite the new material's success, surface scratching was still a major disadvantage compared with the hardness of glass. The Minnesota Mining and Manufacturing Co (3M) were specialists in coating technology and began to address this problem in the early 1970s. They eventually concluded that the key to production of a scratch-resistant coated lens was a dust-free environment more stringent than any they had used previously. After refining their process 3M bought Armolite in 1979, transferring the new technology to them and marketing the coating as RLX Plus.

The Columbia Southern Resin Co had named their find after its batch number, 39. The research project's name was 'Columbia Resin'. Thus was born CR39, the plastic in your spectacle lenses.

A QUESTION OF MURDER

An optical family's links to two notorious murderers

There have been more eminent ophthalmologists, although Nathaniel Bishop Harman achieved much in his own right. But to have two of one's children connected to the defence of two of the twentieth century's most notorious killers is surely unique. In one case, however, the word 'alleged' must be applied, as it was due largely to the testimony of Harman's son that the accused was acquitted of murder.

Nathaniel Bishop Harman was born in 1869, the seventh child but only son to survive, of a well-connected City family. He attended City of London school, and did his medical training at the Middlesex Hospital in addition to taking a double first in Natural Sciences at Cambridge. After volunteering to serve as a surgeon in the Boer War, he returned from South Africa to work at Moorfields. By taking the post of ophthalmic surgeon at the Belgrave Hospital for Children in 1901, he ignited a lifelong interest in reforming the education of children with poor sight. He began other long associations with the West London Hospital and the British Medical Association, and was consultant oculist to the National Institute for the Blind.

Despite contributing to many other ophthalmic bodies, it is his educational campaigning for which he is primarily remembered. His success in getting established special classes and, later, special 'myope' or 'sight-saving' schools for vision-impaired children in

London, gained him international renown, not least in America, as many countries followed the city's example. His ophthalmic legacy includes various devices that he invented, including the 'diaphragm' binocular vision test, a self-illuminating ophthalmoscope, refractometers and scotometers.

In 1905 Nathaniel Bishop Harman married Katharine Chamberlain, herself a remarkable woman. The niece of the politician, Joseph Chamberlain, and cousin of Neville, she was one of the rarer female species for that age, a qualified doctor. They had five children, one of whom died young. Their son, John, and daughter, Elizabeth, are the links in this story.

It was probably never in doubt that John Bishop Harman would go into medicine, given that he was born at his father's practice at 108 Harley Street, in 1907, where John was to live and practice all his life. After Oundle and Cambridge, he completed his medical studies at St Thomas', where he rose to consultant physician rank. One of his interests, clinical pharmacology, was to lead to him being called as a defence witness in the trial of suspected serial killer, John Bodkin Adams.

Adams was an eccentric character, born into a family of Plymouth Brethren in Randalstown, now Northern Ireland, in 1899. He qualified as a doctor without any great distinction and was considered somewhat of a loner. After an unsuccessful stint as a houseman at Bristol Royal Infirmary he took a general practitioner post in Eastbourne where, despite suspicions about his methods, he became very successful in his practice. He also became very wealthy, largely as a result of persuading many of his elderly patients to include him in their wills; who then tended to die soon afterwards. Adams would claim that the high doses of morphine and other drugs that he administered to these patients were for the amelioration of painful symptoms only; and that the bequests he received were in lieu of fees. Others were not so charitable, and rumours about him 'bumping off' patients abounded.

A police investigation focused on two of Adams' patients, both elderly widows: Edith Alice Morrell and Gertrude Hullett. After being partially paralysed by a stroke, Adams treated Morrell with a cocktail of heroin and morphine in order to ease her 'cerebral irritation'. There were several alterations to her will before her death in 1949 – from natural causes according to Adams – that made various gifts to him. The final version removed any bequest, but he still received money, cutlery and a Rolls Royce. Seven years later, Adams prescribed large doses of barbiturates to treat Hullett's depression after her husband's death. She became comatose and eventually died after taking sleeping pills in an apparent suicide attempt. An inquest declared suicide, but the pathologist suspected barbiturate poisoning. Hullett also left her Rolls Royce to Adams, which he sold six days later.

Adams was charged with Morrell's murder, but there was a huge amount of political intrigue around the case. The police were surprised that the Attorney General, Reginald Manningham-Buller, chose this case rather than Hullett's, as Morrell had been cremated, leaving no chance of analysing remains for poisons. He also allowed a crucial police report to be leaked to the defence and, when vital nurses' notebooks went missing from the Director of Public Prosecutions Office, he declined to familiarise himself with them when produced by the defence. Manningham-Buller's fearsome reputation led to him being labelled 'Bullying Manner' but the judge afterwards criticised his weak performance. (His daughter, Eliza, served as Director General of MI5 during 2002-7.) The prosecution's two expert witnesses disagreed while, for the defence, Bishop Harman was convinced that Adams' treatment, while unorthodox, was not reckless. The jury took forty-four minutes to return a not-guilty verdict.

The Home Office pathologist, Francis Camps, originally had examined 310 death certificates issued by Adams, of which 163 were deemed suspicious, often cases where Adams had administered 'special injections'. Camps was convinced Adams was a serial killer. Adams was later convicted of forging prescriptions and offences under the Dangerous Drugs Act. He was fined £2400 plus costs and removed from the Medical Register. He was, though, reinstated in 1961 and continued thereafter as a sole practitioner. He died in 1983, leaving an estate worth over £400,000.

John Bishop Harman's interest in the law was continued through his marriage to a lawyer and through his four daughters, all of whom became solicitors. One of them, Harriet, is the Labour Party's former Deputy Leader. Returning to John's sister, Elizabeth: she met her husband, Francis Pakenham, the future 7th Earl of Longford, at an Oxford summer ball. He spent many years as a Labour politician, but gained notoriety with his campaign in the early 1970s to ban pornography. But he attracted most attention for his efforts to win parole for the Moors murderer, Myra Hindley. His Christian faith, longstanding interest in prisoner reform and liberal outlook had led him to believe that, if society was willing to exercise forgiveness, all offenders eventually could be rehabilitated into that society. Taking on Hindley as his *cause célèbre* was an error of judgement that inevitably tarnished his reputation.

Elizabeth, at one stage, looked set for a fine political career of her own. Eight children proved too much of a barrier, though, so she decided instead to support her husband's political ambitions. Not that she was idle; lured into journalism to write about child-rearing, she then embarked on a stellar career as a historical biographer. Four of her children have followed her literary path, perhaps most notably Lady Antonia Fraser. When Elizabeth died in 2002, aged 96, there were 26 grandchildren and seven great-grandchildren.

Such is the legacy of Nathaniel Bishop Harman, ophthalmologist. There is a neat circular optical connection, though, in that his son, John, was closely involved with the Medical Defence Union (MDU), the world's first medical defence organisation, established in 1885. John Harman was first appointed to the MDU council in 1955-6, and rose to become President in 1976, serving until 1982. And the present Chief Executive of the MDU, Dr Christine Tomkins, is - to close the loop - an ophthalmologist and a Past Master of the Worshipful Company of Spectacle Makers.

My thanks to Seth Belson for the idea, and Christine Tomkins for information on John Harman.

THE OPTICAL CARGO THAT WASN'T

The vital WWII mission almost scuppered by a short-sighted driver

One of the most important 'optical' shipments of the twentieth century contained not a single piece of optical equipment. Instead, the entire consignment comprised a dead body and a briefcase. And the successful delivery of this cargo, of vital importance to a clandestine Second World War plan that might change the whole course of the conflict, was almost fatally compromised by the vanity of a myopic racing driver.

The ingenious British scheme to deceive Hitler was codenamed *Operation Mincemeat*. It was popularised by the film *The Man Who Never Was* (1956), which was based on the book of the same name by one of the protagonists, Lt Cdr Ewen Montague, written in 1953. Montague was tightly constrained by the need for continued secrecy regarding several aspects of the operation, and the film took further liberties with the story. More recently (2010), BBC2 aired a documentary by author Ben Macintyre that filled in more of the facts.

The background to *Operation Mincemeat* was the intended Allied invasion of Italy pencilled in for 1943, following a successful campaign in North Africa. The strategic stepping stone from Africa to southern Europe was obvious: as Churchill said, 'Everyone but a bloody fool would know that it's Sicily.' British intelligence had used successfully techniques of deception against the Germans in North Africa, so MI5 set to work devising a plan to fool the enemy into believing that the Allies would in fact invade the Balkans via Greece in the hope that Axis resources would be diverted from Sicily.

An MI5 officer, Flt Lt Charles Cholmondeley, had previously outlined a scheme whereby a dead body would be dropped into France attached to a defective parachute in order to furnish the Germans with misinformation. He apparently had got the idea from a memo written by another officer, Ian Fleming, the James Bond author. Within MI5, Cholmondeley was on the 'Twenty Committee' which specialised in running double agents and deception. A fellow member was Montague, and together they worked up the idea into a new plan that was to become *Operation Mincemeat*. A body, apparently that of a British officer, the victim of a plane crash, carrying secret strategic documents for an Allied invasion of Greece, would be floated on to the Spanish coast. Spain was officially neutral in the war but in fact largely sympathetic to Germany. Cholmondeley and Montague knew that, on finding the documents, the Spanish authorities would alert German intelligence.

The two rather eccentric MI5 officers set about finding a suitable corpse, which they eventually did thanks to the coroner of St Pancras, London. The 34-year-old homeless man, whose parents were dead and had no other known relatives, had died by ingesting rat poison in a dosage that would in time leave few traces of the cause of death. Many years later he was identified as a Welshman called Glyndwr Michael.

Cholmondeley and Montague then had to create an identity for their fake officer, together with a plausible history and credible supporting documentation that would travel with him in his wallet and uniform. Michael thus became Major William Martin, RM. The all-important document with details of the Greece invasion was to be a letter from Lt Gen Sir Archibald Nye, Vice Chief of the Imperial General Staff, to Gen Sir Harold Alexander, commander of 18th

Army Group, then in Algeria and Tunisia. This, together with other official documents, would travel in a briefcase attached to the belt of the officer's trench coat.

The body was to be delivered to a location offshore near the Spanish town of Huelva by a submarine, *HMS Seraph*, which was at dock in Scotland. This presented the logistical problem of how to get Major Martin to his destination in a suitably believable condition, without too much tissue decomposition. A container was designed by a boffin, Charles Fraser-Smith, who had devised all manner of gadgets for the Special Operations Executive, and was to become the model for Fleming's *Q* in the Bond novels. The canister, six feet six inches long and two feet in diameter, was fashioned from two skins of 22-gauge steel welded together, with asbestos wool lining the inner layer. A similarly-constructed lid closed onto an airtight rubber gasket, sealed with sixteen nuts. The container was to be stocked with dry ice that, on evaporation, would fill the space with carbon dioxide. This would drive out oxygen and allow the body to be preserved without refrigeration. The canister was marked as follows:

'OPTICAL INSTRUMENTS FOR SPECIAL F.O.S.
SHIPMENT'

There was still the problem of how to get the canister to Holy Loch on the west coast of Scotland in double-quick time to rendezvous with *HMS Seraph*. Enter St John Ratcliffe Stewart 'Jock' Horsfall. A motoring enthusiast with no formal engineering training, he loved rebuilding cars for competitive racing. During the First World War his mother was a driver for the head of MI5, and Jock was recruited by them in due course on account of his racing skills. He was the natural choice for the drive to Scotland.

Horsfall selected as his vehicle his own re-engineered Fordson van - used for transporting his Aston Martin - which he claimed he once drove at 100mph down The Mall. He was aware of the cargo he was carrying and, being a practical joker, no doubt enjoyed the thought of deceiving the enemy. One of his japes was to connect up a toilet seat to a battery and wait for his girlfriend to use it. He reported that, 'The scream that Kath gave when the magneto was turned on was most satisfying.' But his eccentricities almost brought the *Mincemeat* mission to a juddering halt. As Macintyre relates, he was quite vain and 'seldom wore leathers or a crash helmet, preferring to race in a shirt and tie.' He was also myopic and astigmatic but preferred not to wear spectacles. On a fast night-time drive in blackout conditions, this was highly dangerous.

Twice on the journey Horsfall narrowly avoided a crash. On the first occasion they had only just collected the canister. Passing a queue waiting at a cinema to see a spy film, they joked that their story was better – and true. Montague recalls in his book, 'And we all burst out laughing to such an extent that Jock almost rammed a tram-standard.' The second incident occurred later on the drive north. According to Montague, 'we could, of course, only use marked headlights; at one point we drove straight across a "roundabout" which, fortunately, only had smooth grass in its centre once we had mounted the kerb.'

Despite the numerous things that could have gone wrong, *Operation Mincemeat* was a success. The body of Major Martin was discovered by Spanish fishermen, the documents were made available to German intelligence, and Hitler acted on the information by moving a whole Panzer division from Sicily to Greece. Consequently the Allied invasion of Sicily, begun on July 10 1943, was achieved in a shorter time and with less loss of life than might have been the case had the ruse not been believed.

The canister of 'optical instruments' did its job perfectly. 'Major Martin' was given a burial with full military honours by the

British Consulate in Spain. But that the cargo very nearly didn't arrive, jeopardising the invasion of Sicily and the possible outcome of the war, was in no small measure due to a driver not wearing his spectacles.

MICKEY, THE EAST END HERO

The optician who saved thousands during the Blitz

The Mickey of the title was Mickey Davis, a three-feet three-inch tall optician often referred to as 'Mickey the Midget' in his local community. He was at the centre of an amazing episode that took place in Spitalfields during the height of the Blitz in 1940. Widely reported at the time, even in America, it is not much known about now, but it is a story that deserves retelling.

When the bombs were raining down on London, the wealthy could retire to safety in the country or, as the government liked to illustrate in a misguided effort to raise morale, were socialising and partying as normal in West End venues (many of which had reinforced basement clubs). The ordinary Londoner had no other option than to stay put, faced with pitifully inadequate provisions for shelter from the air raids. On September 15, 1940, around 100 people pushed their way in to the Savoy hotel demanding shelter; an air raid alert sounded and they refused to leave until the all-clear was heard. They had made their point. Tube stations had become popular places to shelter, but they were not without danger, as bombs hit Balham, Bounds Green and Bank, where 50 people were killed by a direct hit on the ticket hall. Mickey Davis' actions helped to bring about a change in government thinking on protection of the public.

The now-disused Fruit and Wool Exchange, a 1920s art deco building with an imposing facade, stands in Brushfield Street, Spitalfields, in the heart of the East End. During the Blitz thousands of people crowded into its huge basement as people took it upon themselves to find somewhere to shelter. The conditions were horrendous. By 7.30pm, every bit of floor space was covered; people slept on bags of rubbish, there were no sanitary facilities, so the floor became awash with urine, there was no room to move, lighting was dim at best and the passageways were filthy. There was space for an estimated 5000 people but, on the first night of opening, nearly double that number crammed in.

Stepney Borough Council expressed concern at the situation, but it was local activist Mickey Davis, aged 29, who stepped in as unofficial leader of the shelter and set about improving conditions. Mike Brooke, Mickey's nephew and for many years a local journalist, researched the story for the *East London Advertiser* in 2010 in which he tells how Mickey set up first aid and medical units. He not only persuaded off-duty stretcher bearers to tend the sick and injured; he also managed to get a GP acquaintance to make a two-hour journey each day to spend nights in the shelter. Wealthy friends donated drugs and equipment for a dispensary. This free medical service (pre-NHS) was augmented by the introduction of hygiene and disease-prevention practices. Mickey also set up a card index database of all the shelter users. In order to feed the people, he managed to secure donations of food from Marks and Spencer for the setting up and running of a canteen, the profits from which were then used to provide free milk for the children.

Finally the government got around to appointing official Shelter Marshals. This made Mickey redundant, but the first action of the Spitalfields Shelter Committee was to vote the man who had become known as 'the midget with the heart of a giant' back into office as their Shelter Marshal. Mickey's story gained coverage across the Atlantic. Drew Middleton, a reporter for the *New York*

Times, toured the shelter with Mickey and wrote a syndicated piece about him.

Middleton describes Mickey as being less than four feet tall, 'His back humped and misshapen.' The report continues, 'He is master and policeman, judge, father-confessor and elder brother to these thousands of the east end whose homes have been wrecked by the terror that flies by night.' Mickey tells Middleton that, 'He hasn't drawn any pay for it since he moved in on Sept 13 and found "just plain hell." Running the shelter gives him something to do since the Luftwaffe blew his "nice little optician's shop" to bits.' The report was seen across America; for instance in the *St Petersburg Times* (Florida), with the headline on December 1, 1940, 'Dwarf Runs London Air Raid Shelter Which Houses 3,000 Homeless Nightly' and *The Spokesman-Review* (Spokane, Washington) 'Dwarf Master of Refuge' (Dec 2, 1940).

After the war, Mickey continued to live locally, at 103 Commercial Street, where his daughters continued to reside until the 1990s, and continued his public service by becoming a councillor on Stepney Borough Council in 1949. He was elected Deputy Mayor in 1956 but died that year, at the age of 46, before he could take up the post of Mayor which he was due to do the following year. His wife, Doris, who helped Mickey run the shelter, lived on until around ten years ago.

Details of Mickey's optical career are a mystery. The above newspaper report quotes him as saying that his 'nice little optician's shop' was bombed, but details of the shop are elusive. Presumably, the way he describes it, the shop was his rather than him merely working there; but there appears to be no record of the shop's name or location. A trawl of the London trade directories for the period, held at the Guildhall, reveals no businesses in the optical sector in the right area under the name of Davis or Davies. The British Optical Association museum and Worshipful Company of Spectacle Makers archives found no Mickey or Michael Davis in its optical sources for the period and, as his nephew Mike Brooke was only a schoolboy when his uncle died, he has no information about this to pass on. Even the Tower Hamlets Local History Library and Archives, the home of Stepney Borough Council papers, has no reference to Mickey's optical career. Strangely, their file on Davis is listed under 'Davies' with a note that it was often mis-spelt as Davis; Brooke is adamant that the opposite is the case and, as a close relative, one would think he was most likely correct. The Bishopsgate Institute, just around the corner from the Fruit and Wool Exchange, holds a repository of local press cuttings. It may be that lurking there is a report of the bombing of Mickey's shop; but the cuttings are organised by street so finding it would entail either a lucky guess or an extremely time-consuming trawl of the whole archive.

The building which housed 'Mickey's shelter' still survives – but perhaps not for long. When Spitalfields Fruit and Vegetable Market moved out of the area in 1991 there were no local businesses for the wholesalers, who bought their produce from the auctions held at the Fruit and Wool Exchange, to supply. Now the site is up for redevelopment, despite the historic nature of the building and the existence of wartime artefacts and graffiti in the basement. Boris Johnson, Mayor of London, has overruled Tower Hamlets Council's two-time rejection of the development plan. Despite a campaign led by television historian and local resident, Dan Cruickshank, and the Spitalfields Community Group, to get it halted in favour of an alternative plan which would retain the whole building rather than just the facade, a £200 million redevelopment featuring offices and shops will proceed, to be completed in 2018.

It would be a shame if the Fruit and Wool Exchange were to disappear. Mickey Davis certainly deserves to be remembered as an example of selfless public service in the face of official indifference at a time of extreme duress.

My thanks to Mike Brooke for his assistance and advice.

VOYAGE INTO DARKNESS

The horrific tale of an African slave ship

Up from that loathsome prison / The stricken blind ones came; / Below, all had been darkness, / Above, was still the same.

These lines, and others below, are from *The Slave Ships*, by the American Quaker poet John Greenleaf Whittier (1807 – 1892), that retells the gruesome story of the French slave ship, *Rôdeur*, a 200-ton vessel that sailed from the Nigerian port of Bonny on 6 April 1819. Whittier was also an ardent advocate of the abolition of US slavery, and to set the scene of his poem he precedes it with two accounts: one by the French liberal politician and novelist, Benjamin Constant, given in the French Chamber of Deputies on 17 June 1820; the other related in the *Bibliotheque Ophthalmologique* of November 1819.

What follows is a harrowing tale of the destruction wreaked by a highly contagious eye disease in a confined space. Then known as 'Egyptian ophthalmia' after it came to prominence during Napoleon's Egyptian campaign, the disease in question is trachoma. Alternatively known as granulomatous conjunctivitis, it is a chronic inflammation of the mucous membranes of the eyes caused by the parasitic bacterium-like organism *Chlamydia trachomatis*. The resultant scarring leads to corneal opacity and blindness. Modern treatment of early-stage trachoma with antibiotics is largely successful, but in the early nineteenth century there was no cure.

The *Rôdeur* set sail with a crew of 22 fit men and a cargo of 160 healthy slaves. For 15 days, the voyage was uneventful. As the ship neared the equator some of the slaves were brought on deck for inspection and exercise and a marked redness in their eyes was noted but not thought serious. Some fresh air and water was thought to be in order. As the voyage continued and the supply of fresh water dwindled, the slaves' meagre daily ration of eight ounces was cut to half a glass and, allied with the foetid atmosphere in the tightly-packed hold, the symptoms worsened. The ship's surgeon ordered some of the slaves to be brought from below to be given an eyewash made from an infusion of elderflower. Frightened, and expecting torture rather than treatment, the slaves broke free and, clutching each other, leapt overboard in the vain belief of being transported back to their homes.

Most slave ship captains would have immediately disposed of any slave showing signs of illness in the hope of containing an outbreak of contagious disease. The *Rôdeur*'s had at least attempted to effect a cure, but now all efforts were aborted and, according to Constant's account, he had several slaves who had been detained from jumping ship shot or hanged as an example to those remaining. If his attempt to cure the disease was a form of rough kindness, the retention of the other diseased slaves was to prove disastrous to his crew as one by one they succumbed too.

'Overboard with them, shipmates!' / Cutlass and dirk were plied; / Fettered and blind, one after one, / Plunged down the vessel's side.

The sailors' eyes started to itch and their eyelids become swollen and red. As the pain increased they sought relief with the application of hot rice poultices. A thin, yellowish, discharge appeared in the eyes which became thick and greenish as the disease and the pain progressed. Eventually the eyes became swollen shut. Vermicelli was substituted in the poultices as the supply of rice ran out. Exposure to steam vapour was tried by some. Desperate remedies were attempted by others, such as mustard baths or drops of brandy in the eyes to afford some little pain relief. The ship's surgeon, in a futile act of self-surgery, blinded himself.

Even now, the captain calculated the cost of maintaining a blemished, and therefore unsaleable, cargo of slaves. He also knew that the death of slaves by illness or disease was not covered by the ship's insurance, but death by drowning was. Thus he ordered 36 blind slaves to be thrown overboard. Finally, there was but one man among the whole ship's complement who remained with sight, whose seemingly impossible task it was to somehow steer the vessel to safety.

'Help us! For we are stricken / With blindness every one; / Ten days we've floated fearfully, / Unnoting star or sun.'

By great fortune the seeing sailor spotted another ship, whose course the *Rôdeur* eventually passed close enough to for a distress signal to be sent. The crew's relief was cut short when, stunned in disbelief, they heard the reply from the other vessel, a Spanish slave ship, the *Léon*. It was a distress signal of its own as, by terrible coincidence, the *Léon* had been struck down by the same disease and had been drifting with a sightless crew of its own. Unable to help each other, the ships carried on their separate courses, the *Léon* never to be heard of again.

After another twelve days had passed in utter misery and dread, some of the crew found that their condition was easing somewhat and could resume light duties. The others had to endure repeated attacks of the agonising disease. Finally, after almost three months at sea, the *Rôdeur* came to anchor - with no small thanks to the lone sighted crewman - at the island of Guadaloupe on 21 June.

There is a painting by JMW Turner (1840), *The Slave Ship*, or *Slavers Throwing Overboard the Dead and Dying – Typhoon Coming on*. Portraying a similar tragic story to the *Rôdeur's*, Turner's painting was controversial: Mark Twain said that it reminded him of a 'tortoise-shell cat having a fit in a platter of tomatoes'; while John Ruskin, who owned it for many years declared, 'If I were reduced to rest Turner's immortality upon any single work, I should choose this.' The subject is the English slave ship, *Zong*, which departed the west coast of Africa in 1781 bound for Jamaica with 417 slaves. Before long, illness swept the ship and 60 slaves and seven crew were dead. Many of the remaining slaves were sick, and the decision was taken by the captain to shackle together 132 slaves and have them thrown overboard; the insurance would cover the loss.

A group of ten slaves broke free, and leapt overboard. One of these, named Equiano, survived and later settled in England. When he heard of the subsequent court case between the ship's owners and the insurers, who were refusing to pay out, he recounted his tale. The captain's story that the slaves were sacrificed to preserve a dwindling water supply was found to be untrue, not least because the *Zong* reached Jamaica with 420 gallons of water on board. And the moral outrage caused by the legal action (which itself said nothing about the morality of the slaves' treatment) gave a boost to the abolitionist cause.

In the sunny Guadaloupe / A dark-hulled vessel lay, /
With a crew who noted never / The nightfall or the day.

So the nightmare was over for the *Rôdeur*. Fresh food and water were in plentiful supply. Relief from the pain was obtained by application of a simple lotion of lemon juice mixed with spring water which was recommended, ironically, by an inhabitant, a former slavewoman. But the disease had exacted a high toll on the *Rôdeur*. Of the original complement of 22, twelve crew members were totally blind and five, including the captain, had lost the sigh of one eye. Of the slaves who actually reached the island, 39 were totally blind, twelve had sight in only one eye and 14 were reportec to have corneal opacities. And the sole unaffected crew member who had heroically steered the ship to safety? He succumbed to the disease three days after landfall.

2. Art, Culture and Philosophy

THE PINCE-NEZ MYSTERY

Sherlock Holmes finds an elementary optical clue

It would not be too difficult to find references to spectacles or eyesight in literature, but instances of optical appliances providing the pivotal moment of a story seem to be somewhat rare. That Sir Arthur Conan Doyle is responsible for one example should be of little surprise, since at one stage in his medical career he specialised in ophthalmology. Indeed he entitled this particular Sherlock Holmes short story *The Adventure of The Golden Pince-Nez*. There is another example, from the once-popular 1930s 'Dr Thorndyke' series of stories, of a pair of bifocals - most likely Franklin splits (see *The Politician and the Piano,* page 121) - providing an important clue; otherwise, it would appear that the great optical mystery novel is still waiting to be written.

Considering the optical connection under discussion, there is a wonderful irony in the fact that it was Conan Doyle's complete lack of success in his ophthalmology practice that led him to pursue his writing career at the expense of his medical one. He had studied medicine at the University of Edinburgh from 1876 to 1881, gaining experience working in Aston and Sheffield during this time. It was also during this period that he began writing and had his first short stories published. On completing his doctorate in 1885 he became a partner in a medical practice in Plymouth, but the arrangement was not successful; so he left, with very little money to his name, to start his own practice in Southsea, Portsmouth. As might be expected in a brand new practice, initially patients were thin on the ground and Conan Doyle began writing again to fill the time between consultations. His first Sherlock Holmes story, *A Study in Scarlet*, was published in 1887.

Then, in 1890, Conan Doyle decided to study ophthalmology, travelling to Vienna, one of the great centres of the subject at that time, and Paris. On his return to England, he set up an ophthalmology practice in London at Upper Wimpole Street, near Harley Street. This created another opportunity for writing as, according to his autobiography, not a single patient crossed his door. His literary success was assured by the popularity of the Sherlock Holmes stories and he was able to give up medicine to write full- time.

The Adventure of the Golden Pince-Nez was one of the short stories collected in *The Return of Sherlock Holmes*, published in 1904. The tale concerns the murder of a young man who is found clutching 'a golden pince-nez, with two broken ends of black silk cord dangling from the end of it.' His last words were, "It was she!" Since the victim had excellent eyesight, the appliance must have been snatched from the murderer's face. Let Dr Watson take up the story:

'Sherlock Holmes took the glasses into his hand and examined them with the utmost attention and interest. He held them on his nose, endeavoured to read through them, went to the window and stared up the street with them, looked at them most minutely in the full light of the lamp, and finally, with a chuckle, seated himself at the table and wrote a few lines upon a sheet of paper:-

Wanted, a woman of good address, attired like a lady. She has a remarkably thick nose, with eyes that are set close up on either side of it. She has a puckered forehead, a peering expression, and probably rounded shoulders. There are indications that she has had recourse to an optician at least twice during the last few months. As her glasses are of remarkable strength, and as opticians are not very numerous, there should be no difficulty in tracing her.'

How did Holmes come to these conclusions? Let him tell us himself: 'It would be difficult to name any articles which afford a finer field for inference than a pair of glasses, especially so remarkable a pair as these. That they belong to a woman I infer from their delicacy, and also, of course, from the last words of the dying man. As to her being a person of refinement and well dressed, they are, as you perceive, handsomely mounted in solid gold, and it is inconceivable that anyone who wore such glasses could be slatternly in other respects. You will find that the clips are too wide for your nose, showing that the lady's nose was very broad at the base.'

As to the lenses, Holmes explains as follows: 'My own face is a narrow one, and yet I find that I cannot get my eyes into the centre, or near the centre of these glasses. Therefore, the lady's eyes are set very near to the sides of the nose. You will perceive, Watson, that the glasses are concave and of unusual strength. A lady whose vision has been so extremely contracted all her life is sure to have the physical characteristics of such vision, which are seen in the forehead, the eyelids, and the shoulders.'

And the double visit to the optician? Says Holmes, '... the clips are lined with tiny bands of cork to soften the pressure upon the nose. One of these is discoloured and worn to some slight extent, but the other is new. Evidently one has fallen off and been replaced. I should judge that the older of them has not been there more than a few months. They exactly correspond, so I gather the lady went back to the same establishment for the second.' Without wishing to spoil the story's ending, it is safe to say that the murderess's uncorrected myopia, through the loss of her pince-nez, is what leads to her making the error that results in her capture.

R Austin Freeman, a London-born doctor, created his detective, Dr Thorndyke, in homage of Sherlock Holmes. As a barrister and doctor, Thorndyke was versed in law and forensic science. But there is another twist to these stories, for Freeman is credited with what might be called the '*Columbo*-style' of inverted detective story where the crime and the perpetrator are described first. In *Dr Thorndyke Intervenes* (1933), Thorndyke is questioning a witness about a suspect's appearance; having established that he's a spectacle wearer, Thorndyke probes for detail:

'I dunno,' replied Bunter. 'Spectacles is spectacles. I ain't a optician.'

'Some spectacles are large,' said Thorndyke, 'and some are small. Some are round and some are oval, and some have a line across as if they had been cracked. Would his fit any of these descriptions?'

'Why yes, now you come to mention it. They was big round spectacles with a sort of crack across them. But it couldn't have been a crack because it was the same in both eyes. I'd forgotten them before you spoke.'

Thorndyke's Watson-like assistant concludes that 'he had mentioned a very uncommon kind of spectacles – the old-fashioned type of bifocal, which is hardly ever made now ... I had no doubt ... that he was describing a particular pair of spectacles.'

There can be no doubt, though, that Conan Doyle is the premier optical-literary connection, as one of the world's most famous authors. So consider once more Conan Doyle sitting in his consulting room in Upper Wimpole Street, flush with the latest knowledge in ophthalmology. Balancing the conspicuous absence of patients with the money coming in from the Holmes stories, Conan Doyle's making the conclusion that ophthalmology's loss would be literature's gain would be, of course, elementary.

OLD MASTER IN OPTICS

The court painter and his enigmatic subject

In 2012 there was a rare opportunity to view a beautiful inting of an eighteenth century optician's workshop that is not rmally accessible to the public. *John Cuff and his Assistant* by Johan ffany is owned by the Royal Collection and usually hangs in the ivate apartments of Windsor Castle. But from 10 March – 10 June it is on show at the Royal Academy in an exhibition entitled *Johan ffany RA: Society Observed.* Who exactly, though, was John Cuff, d why should such a celebrated artist have painted him?

Cuff was born in 1708 to a watchmaker who was a member of e Broderers' Company. He was apprenticed in 1722 to James Mann, e of a family of optical instrument makers prominent in the orshipful Company of Spectacle Makers (WCSM). Cuff himself was mitted into the freedom of the Company in 1729, giving 40 years' rvice to it, during which time he attended over 100 Court meetings, cluding a stint as Master in 1748.

He set up shop in Fleet Street, in 1737, under the sign of the eflecting Microscope & Spectacles'. As well as optical instruments he ade and sold barometers, thermometers and mathematical and her instruments. One of his innovations was an improved solar lescope. Cuff's big break came when this instrument was described vourably by Henry Baker, an influential Fellow of the Royal Society, in s extremely popular book *The Microscope Made Easy* (1742). The book also described and illustrated some of Cuff's microscopes, making him well-known in the field. In 1743 Cuff developed an all-brass compound microscope with the tube supported by a single pillar mounted on a wooden box foot which was so successful that it was widely copied, and similar designs became known as 'Cuff-type' microscopes.

Although Cuff was a frequent visitor to the nearby Royal Society and had a prominent advocate, and possibly patron, in Baker, he failed to garner sufficient support when put up for election that year. His business acumen was rather lacking, too. In 1750 he was declared bankrupt but managed to continue trading with the proceeds from the sale of his household goods. Another commercial blow was the arrival next door to his premises of the brash, business-savvy optical instrument maker, Benjamin Martin (described in *Maverick of Fleet Street*, see page 125). Baker wrote to a friend, 'one Martin (a Man unknown to me) took a House adjoining to his, and by advertising, and puffing, and by the Mistakes of many who took one Shop for the other, did him much Disservice.'

Martin's willingness to 'borrow' from Cuff's designs and his flair for self-promotion soon drove Cuff away, if only to another part of Fleet Street. For a year or so (1757-8) he ran a shop under the sign of the 'Double Microscope, three Pair of Golden Spectacles & Hadley's Quadrant'. He gave up keeping retail premises after further financial difficulties forced him to auction off his stock but continued to make instruments to order and take on apprentices, latterly at a workshop in the Strand.

At least his pre-eminence as an instrument maker was not in doubt; for instance, there are records showing that payments were made to 'Mr Cuffe' on behalf of George III in both 1770 and 1771 in order to provide him with such materials as a diamond, grinding tools, emery and six chucks for his lathe. His being known to the King was probably how he came into contact with Zoffany, who, incidentally, was similarly poor with money despite his renown as an artist.

Johan Zoffany (1733-1810) was born in Frankfurt and served apprenticeships under artists in Regensburg and Rome before arriving in London in 1760, the year of George III's accession to the English throne. Largely through his acquaintance of the famous actor, David Garrick, and his portraits of him; and the pieces he exhibited at the Society of Artists exhibitions, he came to the notice of the King's advisor, the Earl of Bute. Commissions from Bute finally led to a commission from the King and Queen in 1764 to paint their infant sons. With the patronage of the Royal couple, Zoffany had truly arrived.

Thanks to the support of King George, at whose prompting he was elected a Royal Academician, Zoffany was busier than ever with commissions and was making good money. But his profligate nature meant that he was using it even faster. It seems that the King may have been aware of this, giving him new commissions in order to help ease his financial worries. One of these may have been to paint the portrait of Cuff, although it has been mooted that Zoffany may have offered a portrait to Cuff in lieu of payment for optical services rendered. The consensus view, though, is that it was either painted for or purchased by the King who, as mentioned above, supplied Cuff with funds for materials and probably purchased his microscopes.

The painting, dated 1772, shows the optician seated at a workbench grinding a lens on a treadle-operated lathe. He is wearing protective clothing against glass dust on his head and arms; his spectacles rest above his forehead. His bench and shelves are populated with tools, glass and pots. Behind him stands his assistant. Some questions linger: there is some doubt as to whether the optician is indeed John Cuff; assuming that it is, can his assistant and the workshop location be identified?

According to Oliver Millar's authoritative *The Later Georgian Pictures in the Collection of Her Majesty The Queen* (Phaidon, 1969), the painting was originally exhibited at the Royal Academy in 1772 under the title, *An optician, with his attendant*. It was subsequently described as *Two Old Men*, the sitter being identified variously as 'Mr

Cuff' or 'Mr Cuffs', and later still as *The Lapidaries.* A now-invisible pencilled remark on the stretcher made around 1859 asserted 'Dollond the Optician in the Strand London' which has been taken to refer to Peter Dollond, Optician to the King, who perfected the achromatic telescope. But, as Millar notes, Dollond was born in 1731, surely too young to be Zoffany's subject.

Accepting Cuff as the sitter, then a list of his apprentices is known. The last of these was William Cox, whom Cuff turned over to another optician, Charles Lincoln, in 1771. Assuming that the painting took a year to complete, as Millar says is usual for Zoffany, Cox could be a candidate; but it is hard to imagine that the old-looking figure assisting Cuff is an apprentice. Perhaps he will never be identified with certainty. The workshop's location presumably is the place in the Strand, mentioned above, that he kept from 1764; but this remains to be confirmed.

To gaze at Cuff's image is to see a man who would die the year the painting was completed. At some point he had printed (and possibly wrote himself) *Verses Occasion'd by the Sight of a Chamera Obscura*, which was sold by Mrs Cooke in 1747. The following extract could stand as apt testimony to the craftsmanship and financial tribulations of both Cuff and Zoffany:

Each wondrous work of thine excites Surprize;
And, as at Court some fall, when others rise;
So, if thy magick Pow'r thou deign to shew;
The High are humbled, and advanc'd the Low;

Thanks to Anna Reynolds, Curator of Paintings, The Royal Collection and Colin Eldridge, Archivist and former Clerk of the WCSM for their invaluable help; and Frank Norville for his suggestions and encouragement.

THE NOVELIST, DIARIST AND PRINCESS

Royal and literary connections to a remarkable physician

How are a famous diarist, one of England's great novelists and a princess, destined to become queen of England, connected optically? The man who provides the link is the physician and oculist known as Turberville of Salisbury. He was born at Wayford, near Crewkerne, in Somerset in 1612, but lived and practised medicine for most of his life in Salisbury, hence the appellation.

Dawbigney Turberville was born into an old English family that could trace its heritage back to one Sir Payan D'Urberville, a Norman knight who fought under William the Conqueror at Hastings. Indeed, his name sometimes appears as D'Aubegney D'Urberville. Here is our first connection: the surname in this form was, of course, used by Thomas Hardy for *Tess of the D'Urbervilles*. Whether our man was the inspiration for Hardy can only be surmised, but it is an unusual name and Hardy certainly knew Salisbury and most likely had heard of Turberville, one of her most notable inhabitants. In Hardy's Wessex Salisbury is called Melchester and is where, in *Jude the Obscure*, Jude goes to study for entry into the ministry and where his soon-to-be lover, Sue Bridehead, is training to be a teacher.

When thinking of a famous diarist, one's mind is likely to turn firstly to Samuel Pepys. This is the second connection with Turberville and by far his most well-known, or at least the best-documented, thanks to the Diary in question. That Pepys had problems with his eyes is also common knowledge, since he mentions them many times in his Diary. The first entry concerning them is for 25 April 1662; although this is in relation to the amount of alcohol he has imbibed that day: 'but I was much troubled in my eyes, by reason of the healths I have this day been forced to drink.' The last reference to his eyesight comes in his final Diary entry on 31 May 1669: 'And thus ends all that I doubt I shall ever be able to do with my own eyes in keeping my journal I being not able to do it any longer, having done now so long as to undo my eyes every time I take a pen in my hand.'

Throughout these years of ocular discomfort, which he came to believe would make him blind if he continued writing, he recounts trying many potential remedies. He mentions, in order, being bled, changing his brewer, using green glasses, using 'young glasses' as opposed to 'old spectacles', wearing 'tube' spectacles, a vizard (type of visor) with lenses and the use of a water globe. Yet it was some years before he took the advice of a good friend in getting proper medical attention. In his Diary entry for 22 June 1668 Pepys, then aged 35, writes, 'My business was to meet Mr. Boyle, which I did, and discoursed about my eyes: and he did give me the best advice he could, but refers me to one Turberville of Salisbury, lately come to town, who I will go to.'

Pepys did not waste time in consulting Turberville, who was up in London on one of his regular trips from Salisbury. The following day, writes Pepys, '... to Dr. Turberville about my eyes, whom I met with: and he did discourse, I thought, learnedly about them; and takes time before he did prescribe me anything, to think of it.' On 29 June: 'to Dr. Turberville's, and there did receive some direction for some physic, and also a glass of something to drop into my eyes: he gives me hopes that I may do well.' On 4 July Pepys records taking four of Turberville's pills but, by 13 July, Pepys is seeking another cure: '... this morning I was let blood, and it bled about fifteen ounces, toward curing my eyes.' And by the end of

July he writes, 'The month ends mighty sadly with me. My eyes being now past all use almost; and I am mighty hot on trying the late printed experiment of paper tubes.'

Biographical sources suggest that Pepys never suffered the same level of ocular discomfort after his consultations with Turberville, although his eyes troubled him throughout the rest of his life. A group of ophthalmologists, invited by the Samuel Pepys Club in 1911 to consider the available evidence, hypothesised that Pepys' problem was hypermetropia with some degree of astigmatism. Later on presbyopia would, of course, have become relevant, and who knows what effect trying to read and write in smoky, dimly-candlelit rooms must have had on his visual comfort.

At about the same time as he was treating Pepys, Turberville was attending an even more illustrious patient: the third part of the connection. The Duchess of York, wife of the future James II, was worried about her daughter, Princess Anne. She was suffering from a serious eye inflammation and facial skin eruption which the Court physicians had been unable to cure. Turberville was sent for. The Court physicians protested at the presence of what they considered to be a mere country doctor in their midst and, in the ensuing argument, Turberville refused to communicate with them. The deadlock was broken when the Duke and Duchess of York backed Turberville decisively to the tune of a £600 fee if he took the case on – a huge sum at that time. Unsurprisingly Turberville took up the offer, and successfully treated the future Queen Anne. He only ever received half his fee but, as this was still a substantial amount, the kudos of a happy royal patient was probably sufficient recompense.

Turberville's skills are attested to by his friend, Dr Walter Pope, in his *Life of Seth Ward, Bishop of Salisbury*. Pope says, with obvious gratitude, 'It was he who twice rescued me from blindness, which without his aid had been unavoidable, when both my eyes were so bad, that with the best I could not perceive a letter in a book, not my hand with the other, and grew worse every day.'

Exactly what his problem was we don't know, except that he developed a severe eye infection that led to him resigning his teaching post in 1687. Pope also mentions, to illustrate his integrity, how Turberville once refused a fee of £100 from a peer because he could not cure the patient.

In one of his few correspondences to *Philosophical Transactions*, Turberville mentions his use of a new technique whereby, 'A person of Salisbury had a piece of iron or steel stuck in the iris of the eye [probably, actually at the corneal limbus] which I endeavoured to push it out with a small spatula, but could not; but on applying a lodestone it immediately jumped out.' This very early report of retrieving a foreign body by magnet, together with reports of his performing enucleations and his skill in correcting ptosis, and curing ulcers and inflammations, shows why he earned the reputation of a competent and honourable physician.

The epithet 'of Salisbury' says much about the regard in which Turberville was held. Patients travelled from across England and, indeed, from as far away as Jamaica to consult him. Such was Turberville's reputation, that when he died in 1696, he was afforded burial in the nave of Salisbury Cathedral. His friend Pope supplied a fitting epitaph: 'Near this place, lies interred the most expert and successful oculist that ever was, perhaps that ever will be...'

THE GLASGOW BOY

The artistic legacy of an optical technician

What do opticians keep in the back room of their premises? n times past, it was often the optician who was in the back room when here were such things as jeweller-opticians and chemist-opticians. But vhat about the probably unique case of the electrical repairman-ptician who, in the early twentieth century, used his back room to roduce distinctive paintings whose combined sales have fetched everal hundred thousand pounds.

It's fair to say that art was John Quinton Pringle's (1864 – .925) first love, rather than optics. This artistic tendency may have een inherited from his grandfather, a carpet designer named James hristie. When Christie's son was orphaned, he was looked after by a uardian, John Pringle, whose surname he adopted. He married nother orphan, the couple having eight children – seven boys and a irl – of which John Quinton was the second. Another influence was hat Pringle's father, a railway worker, was appointed stationmaster at angbank, a small town near Glasgow popular with artists. The family noved there in 1869 when Pringle was five years old, spending five ears in an environment in which he would have become used to eeing artists at work.

Two years after moving back to Glasgow's East End, where ringle was to spend almost all his life, he was apprenticed to an ptical repairer and then, in 1878, to an optician, a Mr Davidson, who had a practice in London Road. Nevertheless Pringle's father encouraged his artistic talent – but only so long as it did not interfere with his apprenticeship. In 1883 he was allowed to enrol in the Glasgow School Board's evening classes, with financial assistance from his younger brother, Donald, who maintained his interest in his brother's career, keeping a file of newspaper cuttings of his progress.

Pringle's artwork showed sufficient promise for him to be awarded a bursary to take evening classes at the Glasgow School of Art (GSA), which he attended altogether for ten years. He also took the early morning class there before going on to work. He was a contemporary of the architect, Charles Rennie Mackintosh, and became associated with the group of artists known as the 'Glasgow Boys'. Despite the very conventional syllabus at the GSA, the Glasgow Boys found their influence to a large degree in French realism. Their subjects were naturalistic scenes of rural and everyday Glasgow life, a reaction against the Edinburgh-centred Scottish art establishment. The school's director, Francis Newberry, said, 'The utmost liberty and freedom were allowed; always provided that a man produced something that was his own conception and which bore the stamp of the producer's personality.'

In 1896, a year after finishing at the GSA, Pringle's father died, his mother having passed away six years previously. At around the same time Pringle opened his own optician's practice; or, more accurately, an optical and electrical repair shop. Based at 90 Saltmarket, and fitted out by three of his brothers, a contemporary photograph shows the shop's fascia painted with the words 'Electrician' and 'Optician' either side of his name. There is no record of him having taken any optical qualification, but he became sufficiently well-known in the optical fraternity to merit a brief death notice in *Optician* (01.05.1925), which stated that he 'carried on his business as an optician in Glasgow for many years.... he could never be persuaded to forsake optics for art.'

Douglas Clerk, Pringle's great-great-nephew, who also studied at the GSA, points out in his student dissertation on his bear that it wasn't necessarily a particular love of optics that pped him becoming a professional painter; rather it was the ancial security of owning a business that allowed him to paint and periment at his leisure without the imperative of producing work for e. A fellow student, Muirhead Bone, noted Pringle as being, 'A man o would not dream of becoming a mere professional artist; how en he told us he loved art far too much for such prostitution.' The ck of his shop became his studio in the evenings, outside business urs. Not that he was a great businessman either: he thought nothing closing the shop at short notice for a few days to attend exhibitions, d he could be somewhat careless about collecting payments for rk done.

But he was proud of his craftsmanship and enjoyed his rk, which covered a range of optical, scientific instrumental and ctrical repairs. He was known sometimes to wear two or three pairs spectacles on his head when effecting the most intricate of repairs, d used this mastery of attention to fine detail when painting niatures. As the aforementioned Muirhead Bone later recalled: ere was Pringle, a genius of a watchmaker from the East End, who d the most delicate and fantastic miniatures on ivory and drew "the " in a manner of his own.'

For some years he lived in a flat over the shop with his ter, Mary, who kept house and helped in the shop for almost fifteen ars until her early death at the age of 38. The shock of bereavement d the imminence of the First World War proved to be a difficult time Pringle both personally and business-wise, not least because he w had to run the business single-handedly. In February 1923 he was rsuaded to sell his shop, moving to Invernesshire, at last to paint l-time.

The name over the door in a photograph of the premises of ound 1923-4 is A Schwarz, FBOA. Records show that Anthony hwarz of Glasgow qualified as FBOA in 1913. He was aged 32 in 23, when he bought Pringle's business for £400. The fascia in a later otograph shows the new direction the business took: while retaining the name 'Pringle's', the word 'Electrician' has been replaced by 'Sight Testing'. Schwartz himself must have been an interesting character. He was removed from the BOA register around 1915 and only reinstated in 1919 after paying subscriptions owing dating back to 1916. However after his death in 1948, *Optician* (24.12.1948) noted that 'Many Scottish opticians and all members of the Glasgow & District Local Association are mourning the loss of one of the profession's most enthusiastic and popular members...' so he was evidently well-regarded.

Pringle never quite achieved the fame of some of the other Glasgow Boys, but recognition of his talent while still a student came in the form of several prizes, not least a national award, the 1891 Gold Medal of the Department of Science and Art, South Kensington. He exhibited widely, most notably alongside the likes of Walter Sickert, Wyndham Lewis and Stanley Spencer at an exhibition entitled 'Twentieth Century Art; A Review of Modern Movements' at the Whitechapel Art Gallery. He also achieved the distinction of being recognised within his lifetime with a very successful exhibition of his work at the GSA in 1922. The *Glasgow Herald* reckoned Pringle to be 'in the front rank of living Scottish painters'. There would also be a retrospective exhibition in 1946 at the Saltire Club, and a Centenary Exhibition at the Glasgow Art Gallery, Kelvingrove in 1964.

In 2015 descendants of Anthony Schwartz brought along a painting to an edition of *Antiques Roadshow* held at Kelvingrove. Titled *Portrait of a Girl in a Lace Collar*, it was one of several left in the shop that Pringle said Schwartz could keep. It was valued at £10,000. Better check that back room again!

Thanks to David Pringle and Neil Handley (Curator, British Optical Association Museum) for information; my mother, Elsie Baker (a Glaswegian), for the idea.

LIBERTÉ, ÉGALITÉ, OPTIQUE

The French Revolutionary and scientist who was murdered in his bath

On the one hand he was a forward-thinking physician with a well-to-do clientele, who practised in London and wrote scientific treatises including an account of a hitherto unknown eye disease. On the other hand Jean Paul Marat was a radical at the heart of the French Revolution, who would declaim to readers of his newspaper, 'Five or six hundred heads cut off would have assured your repose, freedom, and happiness.'

Marat was born in 1743 in Baudry, a small town in the Swiss canton of Neuchâtel. The family name of his Sardinian father was actually Mara; Marat added the silent 't' when he moved to France. He had a slight connection to that country through his Swiss mother, who was the descendant of French Huguenots. Marat's radical outlook was evidently established early on. In an autobiographical essay, *A Portrait of the People's Friend, Drawn by Himself*, he wrote: 'I had already developed a sense of morality by the time I was eight years old. At that age, I couldn't bear to see other people treated badly; the sight of cruelty filled me with anger and witnessing an injustice made my heart race as if I myself were the victim.'

He left home at fifteen to further his education, spending five years in Bordeaux and Paris. As was not uncommon at the time, he did not enrol formally at an academic institution, but attended lectures that interested him and read widely. His leanings were towards the sciences and, in 1765, he travelled to England to gain a deeper grounding in those subjects. That he established himself in London as a physician is known, although details of his first five years there are sketchy. Marat is sometimes portrayed as a quack doctor and scientific charlatan, but his medical credentials were in fact sound. He was awarded a medical diploma by the University of St Andrews, signed by two reputable doctors, in 1775. Although he did not attend classes there, it was common at that time to award medical degrees as a sort of certificate of medical competence. Indeed Marat had been practising as a doctor for ten years by then from rooms in then-fashionable Soho, and was soon to publish two distinguished medical pamphlets.

Both pamphlets were written in English and published shortly before he returned to France to practise. The subject of the first, dated November 1775, is an examination of the erupting sores ('gleets') that are associated with gonorrhoea. The second, published in January 1776 describes a previously unknown eye disease that Marat called 'accidental presbytopia' and is entitled *An Enquiry into the Nature, Cause and Cure of a Singular Disease of the Eyes.* This disease, Marat claimed, was often mistaken for *Gutta Serena*, a form of amaurosis – loss of vision in the absence of any apparent lesion. The cause, he wrote, 'is ever the fatal Consequence of taking prepared Mercury without proper care.'

Marat's theory was that physicians were administering harmful doses of mercury-based preparations causing excessive amounts to be released into the bloodstream eventually to act adversely on the eye muscles. The primary symptom was a chronic loss of accommodation, especially noticeable in those too young ordinarily to be presbyopic, hence his term for the condition. Other symptoms included pain on palpation of the eye, difficulty with eye movements and a degree of blur in distance vision.

In his pamphlet Marat proceeds to outline his method of treatment. The rationale for his regimen is predicated on a theory of accommodation whereby it is achieved by the extra-ocular muscles changing the shape of the eye, so the treatment is aimed at 'relaxing, deobstructing and restoring to their due Tone the ocular Muscles.' Firstly, the patient must be put on a very strict, simple diet. Secondly, a routine of weekly bleeding, from the foot, is instigated. The patient is also fed a daily draught of cassia (a cinnamon-like spice) for one month apart from the days bleeding takes place. Then, antispasmodic 'suffumigations' are conveyed to the eyes and emollient poultices applied to the temples. Finally, there is a course of electrical sparks to be drawn from the canthi morning and evening for a few weeks, while a plaster of *tacamahaca* (the resin from a species of poplar tree) is applied to the temples. Then, any remaining weakness of sight can be restored by regular washing with water.

Marat describes three case histories. He admits that these are the only patients he actually treated for the condition, but maintains that their stories demonstrate sufficiently that the treatment is rational and safe. The first subject is an eleven-year-old girl, Charlotte Blondel. A remedy of 'mercurial cakes' for worms resulted in salivation, a swollen head and loss of vision. Further treatments cured all but the latter symptom, which was given up on as incurable by various physicians. On taking over the case, Marat describes how he 'framed a scale' to measure the closest point that Blondel could see the time on a watch. To begin with the measurement is twenty eight inches. This improved to twenty two inches after a month, and when nine inches is achieved the treatment is considered a success.

Marat's other patients are both young men whose treatment for gonorrhoea with 'sublimate corrosive dissolved in spirit of wine' produced similar visual symptoms. It is worth noting that his use of bleeding would have been uncontroversial; it was still commonly used in the nineteenth century and, in France, into the early twentieth century. His use of electrotherapy via localised electrical stimulation of the affected area, however, was novel and grew out of his interest in experimental physics. Marat was certainly going against contemporary wisdom by opposing the use of mercury for the treatment of ocular disease and, concludes a recent study, his overall conception of ophthalmology was quite modern, as being a combination of medicine, physiological optics and ocular surgery.

Marat went on to write treatises on heat and electricity. Sandwiched between these, in 1780, came a dissertation on optics, *Découvertes sur la lumière*, which was devoted primarily to overturning Newton's theory of diffraction. Unfortunately the implacable opposition to his work from the French scientific establishment in the Academy of Sciences seriously damaged his scientific reputation. But by 1789 the French Revolution was in full swing and Marat's publication of a radical political pamphlet, *Offering to the Nation*, initiated his rise to prominence in the social convulsions that followed. He was also now suffering from a painful, chronic and incurable skin disease that led him to spend increasingly long periods soaking in a medicated bath for relief. There Marat was to suffer a fatal stabbing pain, caused by the knife wielded by his assassin, Charlotte Corday, on July 13, 1793.

If Marat's scientific reputation is yet to be rehabilitated, there are witnesses who suggest that at least his revolutionary actions were truly motivated by sympathy for the poorest in society, rather than by a lust for power as perhaps could be said of other figures of the Revolution. An associate of Marat wrote to a friend, 'You must know that Marat lived like a Spartan and that he gave everything he had to those who turned to him for help.' And the novelist Victor Hugo commented, ' Oh, beware, human society: you cannot kill Marat until you have killed the misery of poverty.' A fine epitaph indeed.

THE SQUINTING STATUE

A unique London landmark and its link to a famous Renaissance work

In 1988 a statue, unique in London for an optical reason, was erected at the junction of Fetter Lane and New Fetter Lane in the heart of the City. The subject is John Wilkes, the radical eighteenth century journalist and politician and so-called 'ugliest man in England'. His looks didn't seem to bother him much, since he once claimed he could woo any woman in competition with any man provided he be given a quarter of an hour start on account of his ugliness. The memorial depicts his left convergent squint (turned-in eye), the feature that marks it out from all other London statues.

Wilkes was born on October 17, 1725, the second son of Israel Wilkes, a Clerkenwell distiller. The fact that he was not blessed with good looks and had a prominent squint gave no impediment to him leading a rakish, hedonistic life in his early years. This was largely due to his financial circumstances. His mother was the heiress of a wealthy tanner, and her inheritance paid for good schooling in Hertford and a university education at Leiden in the Netherlands. On his return to England he married Mary Meade, a childhood acquaintance who was nevertheless ten years his senior. She was the only daughter of John Meade, citizen and grocer of London, and inherited his fortune and large estate at Aylesbury.

Wilkes' marriage to Mary was not a happy one. Mary was a devout Presbyterian, staid and reclusive, while Wilkes was inclined – and had the means – to enjoy high society. Their proximity to High Wycombe, near where the infamous Sir Francis Dashwood had his Hellfire Club, didn't help either. After ten years of marriage and the birth of their only daughter, Mary (known as Polly), Wilkes left his wife and took up residence in Westminster, where he became associated with Dashwood and other members of his club, notably Thomas Potter, the Archbishop of Canterbury's son, and the Earl of Sandwich. With the help of Potter and William Pitt the Elder, Wilkes was elected as MP for Aylesbury in 1757. Re-elected in 1761, his hopes of higher office were not realised, and he became a bitter critic of the government.

As a means to launch his attacks Wilkes established a radical newspaper, the *North Briton* in 1762, in which he railed against the Prime Minister, the Earl of Bute, the king, George III, and the royal household. Thus began a period of constant battles with the Establishment, leading to censure, law suits, two duels and Wilkes' eventual flight to Paris. He was eventually expelled from Parliament after being found guilty of allowing his printing press to be used to produce copies of an obscene work, *Essay on Woman*, by his friend Potter. He was also in trouble over a particular issue of *North Briton*, No 45, which was held by Lord North to be 'a false, scandalous and seditious libel'. Having already fought a duel with Lord Talbot, the steward of the royal household, over comments in another issue of *North Briton*, Wilkes found himself challenged by Samuel Martin, an ex-secretary of the Treasury. On November 16, 1763, Wilkes was injured severely in the stomach in the ensuing showdown.

For this, and other reasons, not least his conviction over the printing of the *Essay on Woman*, Wilkes retired to Paris, supported by donations from leading Whigs of £1000 a year. He did not return to London until 1768 when he became involved in an extraordinary series of elections returning him as MP for Middlesex, which the House of Commons refused to recognise. He was later elected an alderman of the City of London, taking part in the struggle between the Corporation of London and the Commons which resulted in the freedom to have parliamentary debates published. After serving as Sheriff of the City and Middlesex, he ultimately gained the office of Lord Mayor of London in 1774.

His later life was largely devoted to the promotion of the widening of suffrage to working men and a fairer redrawing of parliamentary constituencies. In 1776 he put forward unsuccessfully a bill 'for a just and equal representation of the people of England in parliament.' But he was elected in 1779 Chamberlain of the City, a post which kept him in very comfortable circumstances until his death at his Grosvenor Square house on December 26, 1797. His ever faithful daughter Polly, who was always by his side and never married, lived on for only five years more.

Wilkes' statue might be unique to London but some research a few years ago suggested that strabismus was a feature it shared with one of the most famous statues in the world. Shaikh and Leonard-Amodeo, in 'The deviating eyes of Michaelangelo's David' (*J R SocMed. 2005 Feb; 98(2): 75-6)* described the discovery of a left exotropia (turned-out eye) in the classic sculpture. The discovery was made possible by being able to analyse a three-dimensional computer model created in 1999 by Stanford University professor of computer science and engineering, Marc Levoy. His painstaking digital reconstruction of *David*, the result of the *Digital Michaelangelo Project*, allows the statue to be rotated and viewed from any angle.

Given that, at a height of five metres plus the pedestal, a direct frontal examination of *David*'s face is not possible from the normal observer's point of view, the computer model provides a completely new perspective. Shaikh and Leonard-Amodeo claim that Michaelangelo knew exactly what he was doing in incorporating purposely the exotropia when sculpting *David*'s eyes. An observer walking around the statue sees only one eye at a time:

from the left-hand side, the left eye appears to the viewer to be looking towards and above him as if *David* is focusing on Goliath, the right eye being mostly hidden behind *David*s sling; moving around the right-hand side, the right eye remains visible while the left eye disappears. The contention is that the appearance of each eye is in accord with the asymmetry of the two sides of the statue. The left side is dynamic, indicating movement in that direction, the left eye searching for the sling's intended target. The right side is smooth and indicative of power and intelligence, and the right eye accords with those qualities.

Levoy suspects that Michaelangelo's clever use of perspective, in giving each eye an optimum appearance from the side, had gone unnoticed for so long because of another, more prominent, feature of the naked *David*s anatomy attracting most of the attention. However one art critic, Adrian Searle of *The Guardian*, cast doubt on the whole premise of this new theory of *David*s eyes, saying it was 'another ridiculous and irritating example of scientists second-guessing artists' intentions when they really don't know what were the conditions at the time.' For good measure he commented, 'It is so easy to do this 500 years later and come up with easy explanations. Particularly when you are working with marble, which is full of imperfections, cracks and subtle variations in density, you just cannot assume these things.'

None of which alters the fact that *David* does indeed appear to have a divergent squint whether intentional or not. So John Wilkes' statue at least has an illustrious Italian connection. Not that he is the only London statue exhibiting some form of visual impairment: remember, there is a rather famous one-eyed admiral who perches atop a Corinthian column 170 feet above the ground in Trafalgar Square.

THE PHILOSOPHER'S GRINDSTONE

Baruch Spinoza and the career that may have killed him

Baruch Spinoza (1632 – 1677) made, in his relatively short life, original contributions to philosophy and theology. He was denounced as a heretic by his Amsterdam Jewish community and then managed to alienate most of the other religious authorities with his philosophy of 'God-as-nature.' He also made very fine optical lenses over a 20-year period.

Biographies of Spinoza have often stated that his lensmaking was merely a sideline that earned him a living. A more recent biographer, Gullan-Whur (*Within Reason: A Life of Spinoza*, 1988), points out that he would not have disputed on optical matters at length in his correspondence or written a *Treatise on the Rainbow* (possibly burned later by him, as unsatisfactory) had his interest in optics not been genuine. He was, like many others in his day, a competent amateur.

What fired Spinoza's interest in optics? He could have studied at Leiden University, a lively research institute at that time. He declined, but would have heard the academic talk of the 'perspective glasse' also called the 'optic magnifying glasse' that had been developed into an 'optic tube' for viewing heavenly bodies. He possessed an old textbook, Scheiner's *Astronomy and Refraction*. He also owned *De Vero Telescopii Inventore*, by Borelius, which explained how Galileo's 'imperfect instrument' with an inverted image had been improved by Scheiner: a convex lens was first substituted for Galileo's concave one, and a second convex lens righted the image. It was at Leiden that Snell first formulated his Law of Refraction, currently being revised there by Huygens, with whom Spinoza was to enjoy a lengthy dialogue. For Spinoza it was more interesting how lenses enabled an accurate view of objects than what objects were viewable.

Spinoza began his optical investigations with mathematical theory, calculating the optimal angle at which a ground lens would refract parallel rays to a focus. Before launching on experiment, he calculated whether it was preferable to purchase tools for making plano-convex or concavo-convex lenses. In a letter, he wrote, 'I have a mind to get new tools made for me for polishing glasses.......I do not see what advantage we obtain by polishing convex-concave glasses. On the contrary, convex-plane lenses must be more useful, if I have made the calculations correctly.....the reason why convex-concave glasses are less satisfactory is that, besides requiring double the labour and expense, the rays, since they are not all directed towards one and the same point, never fall perpendicularly on the concave surface.'

There were four stages in contemporary lensmaking:

1. A blank was sawn from raw slab glass using a hacksaw blade wiped with ash and oil mixed with diamond grit (worked in by sawing on hard glass or quartz).
2. The lens was held in place with black pitch while the surface was rough-ground to size and radius with a centring spindle and lathe (or trepanning mill).
3. Next, fine grinding of the lens while maintaining the radii.
4. The lens was finished with carborundum and polishing powders.

The tools and materials required were readily available in Spinoza's Amsterdam. In 1661 he was observed 'to occupy himself with the construction of telescopes and microscopes' and he must have been learning those skills for some time. He later owned two textbooks on glass-cutting and optics: *Optica Promota*, by John Gregory (1663) and *Art of Glass Cutting* by Antonio Neri (1668).

It is thought that he ultimately added nothing original to optical theory despite his intense studies of refraction and the quality of lenses he produced, which sold 'pretty dear'. Nevertheless he was visited often by Huygens, who took note of Spinoza's theories and methods of lens production - while carefully guarding his own. Despite his secretiveness and academic competitiveness – he charged his brother with keeping tabs on Spinoza's progress while he was away in Paris – Huygens maintained a grudging respect for him, and even allowed him one genuine scientific finding: 'It is true that experience confirms what is said by Spinoza, namely that the small objectives in the microscope represent the objects much more finely than the large ones.'

Leibnitz paid him a visit too. He considered Spinoza an outstanding microscopist, yet not among 'the best observers of our day.' Spinoza even managed to have an argument, via his extensive communications with Henry Oldenburg, Secretary of the Royal Society, with Robert Boyle over the best way to polish lenses. Spinoza asserted that hand polishing was superior to using a lathe, of which his acquisition of the Neri textbook further convinced him. The practicalities of working in confined spaces, as Spinoza always did, may have influenced him too.

Spinoza died aged 44, it is said, from a form of tuberculosis. But his work would have exposed him to powdered glass dust for many years; and perhaps an underlying silicosis was a factor in his relatively early demise. At his death his toolbox held 'some glasses in poor condition, among them one good one, and a small quantity of glass and tin tubing.'

MR MOLYNEUX'S PROBLEM

A famous conundrum of visual perception – and its solution

William Molyneux had a problem. Or at least his scientific and philosophical investigations led him to propose the conundrum that now bears his name. Molyneux's thought-experiment (as it could only have been in his time) addresses the very nature of visual perception. It has engaged many of the finest thinkers from the seventeenth century onwards, notably John Locke and Bishop George Berkeley, but also Leibnitz, Voltaire and Diderot to name just a few.

The background to Molyneux's eponymous query was both his wide scientific studies and an episode of personal tragedy. He was born in Dublin in 1656 to a wealthy English Protestant landowner. An aptitude for science and mathematics was amply demonstrated by the time he graduated from the city's Trinity College aged seventeen. One of his interests was optics; he struck up a friendship with the Astronomer Royal, John Flamsteed, to whom he complained that good optical instruments were hard to come by in Ireland. And one of Molyneux's best-known works is his optical treatise, published in 1692, *Dioptrica Nova*. The previous year his wife had died, after only thirteen years of marriage. Lucy, 'a lady noted for intelligence, amiability and great beauty' according to one biographer of Molyneux, was the daughter of the Attorney-General of Ireland, Sir William Domville. Within two months of their wedding Lucy was taken ill and lost her sight, the doctors being able to find neither cause nor cure. Despite suffering in pain throughout the rest of her short life, the couple had three children, although only one survived childhood.

Molyneux, an admirer of Locke, had read his *Essay Concerning Human Understanding* (1690) and was impressed by its discussion of ideas that we acquire by one sense and those acquired by a combination of senses. He may also have come across the twelfth century philosophical novel, *Hayy ibn Yaqhdan*, by Ibn Tufail, published in an English translation only a few years before (1674) which explores some of the same sensory themes. The combination of thinking deeply about these ideas of sense perception and the ever-present trauma of his wife's blindness led to Molyneux posing, in his own words, a 'jocose problem' to his friend, Locke, in a letter of July 7, 1688 (an extract of Locke's *Essay* had appeared in a French journal that year).

The problem, simply put, is this: a person, blind from birth, has learned to distinguish between a cube and a sphere by touch. On recovering their sight, would they then be able to distinguish between these objects without first touching them?

The answer to the problem depends on one's philosophical point of view. In the first place, in the late seventeenth century it was thought to be impossible that someone blind from birth could latterly acquire sight anyway, so the question was deemed to be entirely theoretical. Empiricists like Locke, Molyneux and Berkeley (who considered this conundrum in depth) held that knowledge was gained from experiences, so their answer was 'no'. Locke and Molyneux believed that there was, however, a direct relationship between the visual and tactile senses, so that the newly-sighted person, on touching the objects, would necessarily connect their previous tactile experience with the visual one and almost immediately be able to learn the visual aspect of the shapes. Berkeley went further, arguing that the relation between the visual and tactile senses was arbitrary so that even this had to be learned by experience, the newly-sighted person having to 'work out' the sensory relationships of the objects.

On the other hand, rationalists like Leibniz believed that - to simplify greatly – some of our knowledge about the world is innate and some can be deduced or inferred. By their understanding, the answer to Molyneux's problem is 'yes'; the newly-blind person will be able to 'know' or infer by reason which object is which, since the visual and tactile notions of an object are essentially the same.

The Molyneux Problem gained a wide audience as a result of Locke including a version of it in a revised edition of his *Essay*. In addition, in 1709 Berkeley published his *Essay Towards a New Theory of Vision*, an empirical treatment of the perception of distance, magnitude and figure, in which he refers to Molyneux's question several times to illustrate various consequences of his thesis. Berkeley's extreme form of empiricism holds that all qualities attributed to objects are sense qualities, and only exist while they are being perceived. This is summed up by his famous dictum, *esse est percipi* (to be is to be perceived).

The English surgeon, William Cheselden caused a stir when, in 1728, he provided the first experimental data relating to the Molyneux Problem by publishing an account of a congenitally blind subject's visual experience after he had had his cataracts removed. Cheselden's account stated that the boy initially couldn't recognise different objects regardless of shape or size. Some took this as confirmation of the empiricists' argument, while others found Cheselden's account to be flawed in several ways, thus invalidating its claims.

Throughout the nineteenth century more experimental evidence was acquired; even if its significance was hotly debated. It largely comprised new accounts of vision after the removal of congenital cataracts, with ophthalmologists experimenting with subjects' perception of form, size and distance. Some, such as Franz and Nunneley actually tried reproducing the original scenario of sphere and cube. Evidential outcomes were claimed for both sides of the debate but, in reality, differences such as varying pre- and post-operative circumstances made results difficult to compare and ultimately inconclusive. Various neurobiological experiments have been proposed in modern times using both animals and humans, to examine the nature of acquired vision (say via visual deprivation of newborn animals) and the relationship among the senses (for example in the use of sensory substitution devices where vision has been lost). All are of interest, but not directly related, to Molyneux's specific question.

The problem of acquiring adequate experimental evidence is finding appropriate subjects. As we have seen, restoring sight to the congenitally blind was for a long time impossible. Now, in the developed world, congenital causes of blindness tend to be diagnosed and treated at a very young age. In 2003, Pawan Sinha, professor of vision and computational neuroscience at MIT, and his colleagues liaised with a medical charity to find five subjects in rural north India, who were congenitally blind (light perception at best) with treatable disorders, but crucially aged between eight and 17. Post-operatively vision was at least 6/48 (just better than seeing the top letter on a normal eyetest chart) in all cases, good enough to perform tests within 48 hours of having vision restored. The children were presented with 20 blocks which they could either see or touch but not both. Almost immediately they could distinguish the blocks by sight alone, as well as by touch alone. But they could barely match a seen block with a felt one. Yet learning was rapid: within days one subject was proficient in matching blocks and, within three months, the average success in matching seen with felt blocks was 80 per cent. This result seems to vindicate Molyneux's and Locke's position.

Molyneux's political writings have echoed down the ages too. He lived through turbulent times, not least the war in Ireland that was part of the 'Glorious Revolution' of 1688. His defence of the independence of the Irish parliament in *The Case of Ireland's Being Bound by Acts of Parliament in England, Stated* (1698) is seen as the first stating of the argument 'no taxation without representation'. That certainly became a problem for the English Crown.

VISION ON SCREEN

Optical drama in film and television

Doctors, whether mad, bad or heroic have been staple characters in film and television almost since the beginning of movies. Dentists can point to Laurence Olivier as a sadistic Nazi dentist in *Marathon Man* (1976) and even an eponymous film, the 1960 comedy, *Dentist in the Chair*, starring Bob Monkhouse. But what of the optical profession on celluloid or video? There may not be a large canon of screen productions that could be entitled 'Optical Movie Heroes and Heroines', but there have been a few films that heavily feature ophthalmologists, one an optometrist, and several where vision or blindness play a prominent part in the plot. What follows is a chronological summary of those films.

The earliest instance of an eye condition being critical to the plot of a film occurs in the fourth of a series of ten Dr Kildare stories, with Kildare played by Lew Ayres in all but the first and his mentor, Dr Gillespie, by Lionel Barrymore. In *The Secret of Dr Kildare* (1939) the good doctor is having to care for an overworked Dr Gillespie, while dealing with the case of a millionaire's daughter who he surmises is suffering from hysterical blindness. He decides to treat the girl by carrying out an elaborate charade of pretending to perform a delicate operation on her eyes. Needless to say he is successful and the girl can see again.

A little-known British crime film of 1939, *Shadowed Eyes*, features a noted eye surgeon who, during a mental blackout, murders his wife's lover. After another bout of amnesia he manages successfully to operate on the man who helped convict him of the crime. Another cure of failing eyesight features in the 1952 film, *This Woman Is Dangerous*. Despite the arresting title and having as its star Joan Crawford, it was not a great success. It was Crawford's last film under her Warner Brothers contract and she complained that 'Nothing was right about "This Woman is Dangerous," a shoddy story, a cliché script and no direction to speak of.' The plot casts Crawford as a master criminal and the girlfriend of a vicious gangster. She learns that she is going blind and seeks out the one doctor who may be able to help. In curing her the surgeon also helps her to have a change of heart about her lifestyle and they fall in love. The gangster boyfriend, having been tipped off about the affair, attempts to do his worst but is intercepted by the FBI.

Continuing the theme of loss of vision somewhat improbably restored, *Magnificent Obsession* is a 1954 film version of Lloyd C Douglas' 1929 novel. Reckless playboy, Bob Merrick (Rock Hudson), is indirectly responsible for the death of a well-loved philanthropic local doctor. The doctor's widow (Jane Wyman) is unimpressed with his ham-fisted attempts to apologise, so he decides to make amends by going to medical school and devoting himself to helping others. Meanwhile the widow loses her sight and for some months is under the care of a brilliant surgeon, with whom she falls in love. Of course, the surgeon is Merrick; through his skill her sight is restored and he is forgiven.

In complete contrast, the 1963 film, *The Man With X-Ray Eyes* concerns a scientist, Dr James Xavier (Ray Milland), whose vision researches lead him to develop a drug that he hopes will enhance human vision to a breadth much wider than the normal visual spectrum. In order to gain objective evidence he tries the eyedrops on himself. Initially it proves useful as he becomes able to see inside the human body and thereby manages to save the life of a young girl whose condition has been misdiagnosed. However as his abilities increase he becomes disturbed by the sheer amount of visual information pouring into his brain. After accidentally killing his colleague, he goes on the run to a carnival and then the Las Vegas casinos. Finally he encounters a preacher at a religious revivalist meeting, to whom he confides that he can now see to the edge of the universe. The preacher declares that he is seeing 'sin and the devil' and quotes from Matthew, Chapter V: 'If thine eye offends thee....pluck it out!' Xavier takes the biblical advice rather than continue his suffering.

An equally – if not even more – grisly offering is *Mansion of the Doomed* (1977). Those growing up in the Sixties and Seventies may remember *Voyage to the Bottom of the Sea*. It featured Richard Basehart as the admiral. Here he plays Dr Leonard Chaney, an eye surgeon who is wracked with guilt after his daughter, Nancy, loses her sight in a car accident for which he is responsible. He kidnaps Nancy's fiancé and harvests his eyes for transplant and the unfortunate donor is consigned, empty-socketed, to the basement until he can be restored some time in the indefinite future. The operation is partially successful, but only for a while; after which a succession of unwilling eye donors are snared and left to languish in the basement.....until finally they escape.

Probably the most famous film to feature an optical professional is Woody Allen's *Crimes and Misdemeanors* (1989). Apart from Allen, the cast includes Alan Alda, Mia Farrow, Anjelica Huston, Claire Bloom, Sam Waterston and Martin Landau as an ophthalmologist, Judah Rosenthal. There are two plot strands, one concerning Allen's character and the other revolving around Landau's. Rosenthal, an esteemed surgeon and respectable family man has, in fact, been having an affair for many years. When his mistress realises that he is never going to leave his wife, she threatens to reveal the affair to her. Rosenthal reluctantly engages his brother to hire a hitman to kill his mistress in order to preserve his marriage and reputation. Eventually the murder is blamed on a drifter, but Rosenthal has to carry the guilt.

An optometrist is the hero of a 1993 TV-movie, *Four Eyes and Six Guns*. In this comedy Western, Judge Reinhold plays the big-city optometrist, Ernest Albright, who moves to the frontier town of Tombstone, Arizona. There he meets Wyatt Earp, but is disappointed to find that he is a dissolute drunk, and myopic too. Good old Albright helps to straighten out Sheriff Earp, who then proceeds to bring order to the town.

It should be obvious by now: but beware novel surgical procedures in movies; they are asking for trouble. In *Blink* (1994), musician Emma Brody (Madeleine Stowe), who has been blind for twenty years, receives a double corneal transplant, performed by an admiring doctor. However the new technique used causes flashes of vision that she cannot place accurately in time. In suffering from this 'retroactive vision' she is unsure whether she has seen a murderer fleeing from the scene of a crime. Although her credibility as a witness is questionable, she is believed by a detective on the trail of a serial killer, who in turn begins to stalk Brody...

Finally, anyone who has seen *Friends* on television will no doubt remember Tom Selleck's appearance as Monica's boyfriend. Selleck (*Magnum, PI*) played Dr Richard Burke in ten episodes, an ophthalmologist twenty-one years Monica's senior and a close friend of her parents.

The subject of eyes and vision seems to lend itself to the melodramatic and gory on screen, perhaps because of people's squeamishness about eyes and fear of losing one's eyesight. Hopefully the Tom Selleck version of the optical practitioner ultimately will lodge more firmly in memories than the Richard Basehart or Ray Milland ones.

PLAY IT AGAIN, EDDIE

The curious - and macabre - background to a well-known pop song

I'm gonna break out of the city, leave the people here behind,
Searching for adventure, it's the kind of life to find.
Tired of doing day jobs, with no thanks for what I do,
I know I must be someone, now I'm gonna find out who.

These are the opening lines of a 1977 hit record. Reaching the top ten of the UK singles chart, it is instantly recognisable to anyone who was a teenager at that time, and is still played frequently enough to be familiar to music radio station listeners. The song and its performers may have influenced many other bands from the punk era onwards, but its relevance here is that it is possibly unique as a chart single in mentioning opticians in its lyrics.

The group responsible for the song was a four-piece band that was formed in Southend-on-sea, Essex, in 1975. There was a fifth 'member' of the band, a dummy called Eddie, who used to appear on stage with them; and he gave his name to the outfit, *Eddie and the Hot Rods*. They developed a cult following on the pub circuit playing hard, fast rock and roll, following the path stomped by their near neighbours from Canvey Island, *Dr Feelgood*. Their style, somewhat more raucous - and more popular - than some of their contemporaries, helped to create an opening for the punk movement in the UK by persuading venues to book acts such as the *Sex Pistols* and *The Damned*. It was also to be their downfall because, as punk took off, the Hot Rods' brand of pub rock came to seem outdated.

Why don't you ask them what they expect from you?
Why don't you tell them what you're gonna do?
You get so lonely, maybe it's better that way,
It ain't you only, you got something to say.
Do anything you wanna do,
Do anything you wanna do.

The addition of guitarist Graeme Douglas to the Hot Rods' line-up (from fellow rockers *The Kursaal Flyers*) seemed to provide the spur for their breakthrough into the singles chart. Signed to Island Records, their first hit, *Teenage Depression*, reached number 35 towards the end of 1976. There was a moderately successful album of the same name, and the *New Musical Express'* February 1977 prediction of the Hot Rods as being the 'Most Promising Emergent Act' appeared prescient as the band reached a high point of number nine in the singles chart that August with the single from which the lyrics quoted above are taken: *Do Anything You Wanna Do*. In addition to breaching the Top Ten, it was named single of the week in much of the music press.

The music was composed by the aforementioned Douglas, while the lyrics were written by their manager and erstwhile Southend DJ, Ed Hollis (the elder brother of *Talk Talk* lead singer, Mark Hollis). The lyrics were a nod to the work of occultist, magician and all-round bad boy, Aleister Crowley. Born in 1875 in Leamington Spa to members of the puritanical Plymouth Brethren sect as Edward Alexander Crowley, he revelled in the moniker applied to him as 'the wickedest man in the world' thanks to his debauched lifestyle. In a work entitled *The Book of the Law* he set out his philosophy of '"Do what thou wilt" shall be the whole of the

.aw.' (It is arguable that the wickedeset thing he did was to name his children Lola Zaza, Aleister Atatürk and Nuit Ma Ahathoor Hecate Sappho Jezebel Lilith.)

Do Anything You Wanna Do was marketed with a record sleeve featuring an image of the bald Crowley with superimposed Mickey Mouse ears. What was intended as a joke ended up causing the band a certain amount of grief. As The Hot Rods' bassist, and later member of *The Damned*, Paul Gray, relates in his online memoir at his website, *paulgraybass website*:

'It wasn't long afterwards that what was to become known as the Curse of the Hotrods struck. In retrospect it wasn't the best of ideas to mess about with Aleister Crowley ...... It wasn't long before the letters started coming from his followers, saying we were playing with fire and threatening dire retributions on us all. At the time it was unnerving and we tried to laugh it off, but uncannily enough we suffered our fair share of tragedies soon after. In no particular order one of the guys responsible for the cover committed suicide, our manager died of a drug overdose and all other sorts of trouble befell band members that I won't go into here.'

Despite a subsequent huge tour of America - 57 gigs in 52 days - and recording sessions enlivened by guest appearances from Jools Holland, Linda McCartney and others, the Hot Rods' time had passed now that the punk movement had taken hold. They parted company and disbanded in 1981. But that was not the end of the story: a brief reprise of the band soon after was followed by another reforming in 1985 and a reunion of the original successful line-up in 1992. Indeed Eddie and the Hot Rods are still touring now, but with just the one original member, vocalist Barrie Masters.

If anyone knows of another track that mentions opticians, let them come forward. But it is humbly suggested that *Do Anything You Wanna Do*, this rousing song with its optical reference, should become a standard at any self-respecting student optical ball. All together now (with apologies for the grammar):

I don't need no politicians to tell me things I shouldn't be,
*Neither no **opticians** to tell me what I oughta see...*

...Do anything you wanna do,
Do anything you wanna do

Postscript

After this article was published in *Optician* a reader contacted the author, pointing out that punk band *The Vibrators* had a track on their 1977 debut album, *Pure Mania*, called *Stiff Little Fingers* (from which the eponymous band formed the same year took its name) that included the lines: 'The optician said 20/20; But you stumbled round like you were blind.' The track was only ever released as a B-side to a single so, although it is another genuine example of a song mentioning opticians, *Eddie's* song remains unique (apparently) as a single.

3. People and Technology

A BUTCHER'S BEQUEST

How a remote French village became a laboratory for genetic research

Jean Nougaret (1637 – 1719) was an ordinary Frenchman, a country butcher from Provence. Yet not quite so ordinary as it turned out, for a reason that was to have important consequences for his family and the little village of Vendemian that he settled in.

Vendemian, located about fifteen miles from Montpellier in the southern Languedoc Roussillon region, lies nestled in a landscape whose slopes and plains are perfect for viticulture. Into this secluded wine-growing community came Nougaret, 'la Provencal', as he became known. It seems he married, had children and worked at his trade without incident until his death. By 1907 his descendants numbered 2076. How do we know this? Genetic research has shown that he was the progenitor of a form of congenital stationary night blindness (CSNB) that has persisted among the inhabitants of Vendemian for at least twelve generations. It is not known, though, whether Nougaret himself came from a branch of an earlier affected Provencal family.

CSNB is characterised by normal daytime vision and reduced, to varying degrees, vision in low ambient light levels. It is a non-progressive (hence 'stationary') condition that can be x-linked (mainly males affected), autosomal recesive or autosomal dominant. The 'Nougaret' form, as it is now known, is of the latter type:

parents are equally likely to transmit the disorder and offspring are equally likely to inherit it. In other words, if one parent has the defective gene, a child has a 50 per cent chance of inheriting the defect.

The outside world first became aware of the Nougaret family in 1831 when one Pierre Mirebagues, a sixteen-year-old French army conscript and fifth-generation descendant of Jean Nougaret, attempted to claim exemption from army service on account of his poor night vision. He was suspected of malingering when the examining medical officer found that he could read by candlelight, and was promptly enrolled for seven years' service. Fortunately for him, he was later re-examined and exempted from army service.

Mirebagues claimed that his father, grandfather and great-grandfather all had the same affliction. Intrigued, the Military Surgeon of Montpellier, M Gastè, told the story to a Belgian ophthalmologist acquaintance, Florent Cunier, who went to Vendemian to interview as many of the afflicted individuals as he could find there. He arranged for a local antiquary, M Chauvet, to compile a genealogy of these inhabitants. By making extensive use of the parish records and village archives, and by talking to the villagers and collating the oral and written traditions of the village, Chauvet was able to establish Nougaret as the common ancestor of all the CSNB sufferers in the vicinity. His genealogy ran to 629 people, of whom 86 suffered from CSNB.

Although Cunier published his findings in 1838 and these bought the Nougaret story to a wider audience, the journal in which he published was not that easily accessible. This, however, had the happy consequence of bringing together a triumvirate who were to join forces to investigate the case in more depth. At the turn of the twentieth century the English ophthalmologist, Edward Nettleship, who was much concerned with inherited eye diseases, became interested in the Nougaret findings. But to obtain a copy of Cunier's paper Nettleship had to apply to Ernst Fuchs, the famous Viennese ophthalmologist, as none was available elsewhere. After making his translation from the original French Nettleship enlisted Dr H Truc, Professor of Ophthalmology at the University of Montpellier, to help with his investigation. Truc, in turn, asked M l'Abbè Alphonse Capion, Curè of Vendemian, for assistance.

Abbè Capion was to assemble a modern genealogy, a task that took two years, eventually producing the largest pedigree for any inheritance of any human condition. There were identified 2076 descendants of Nougaret, of which 135 were known to have or have had night-blindness; the latter comprised 72 males, 62 females and one sex unknown. Truc and Nettleship concerned themselves with examining the external and internal appearance of the eyes of the living night-blindness sufferers. Their ophthalmoscopic and visual field examinations showed all cases to be of CSNB as opposed to progressive retinitis pigmentosa (RP) or the result of any other disease. Nettleship's conclusions were in accordance with Cunier's and Chauvet's, and he published his findings with due acknowledgement to Abbè Capion in *Trans. Ophthal. Soc. UK* (27: 269 – 293) in 1907. Nettleship also became noted for his important contributions to research into ocular albinism and RP; ocular albinism Type I is known as Nettleship-Falls syndrome, and the Nettleship Medal of the Ophthalmological Society was named in his honour.

There were other subsequent attempts to update previous studies of the Nougaret CSNB, notably by Dejean in 1949. Unfortunately, by then the effects of two World Wars and the introduction of modern modes of transport had resulted in the opening up of the Vendemian district and greater movement of people in and out of the region. With Vendemian no longer being such an isolated rural enclave it became much harder to trace Nougaret's later descendants.

Modern advances in genetic science have shifted the emphasis of research into the Nougaret CSNB. Recent research has tended to look at the mechanism of the condition. For the technically-minded, in 1996 Dryja *et al* reported in *Nature Genetics* (Missuse Mutation in the Gene Encoding the Alpha Subunit of Rod Transduction in the Nougaret Form of Congenital Stationary Night Blindness, 13: 358 – 360) that the condition is caused by a missense mutation in the gene encoding the α-subunit of rod transducin – the G-protein that couples rhodopsin to cGMP-phosphodiesterase in the phototransduction cascade. Thus rod transducin joins rhodopsin and the α-subunit of rod c-GMP phosphodiesterase to become the third unit of the rod phototransduction cascade where a defect is implicated as the cause of CSNB.

A 1998 study by Sandberg *et al* (Rod and Cone Function in the Nougaret Form of Congenital Stationary Night Blindness, *Arch. Ophthalmol.* 116: 867 – 872) used sophisticated electrophysiological techniques to analyse two twelfth-generation members of the Nougaret family. Their findings suggest that rod function in this form of CSNB is not completely absent as had been previously thought.

More recently still, in 2006, the biochemical mechanism of Nougaret CSNB was investigated via the study of a specially-created transgenic 'Nougaret mouse'. The details are to be found in the paper of Moussaif *et al* (Phototransduction in a Transgenic Mouse Model of Nougaret Night Blindness, *Neurosci* 26(25): 6863 – 6872). The mice were subjected to biochemical, electrophysiological and vision-dependent behavioural analysis. In summary, these mice were subjected to periods of light and dark adaptation during which their behaviour was monitored and after which their retinas were extracted and analysed. Two key deficiencies in the defective gene function were found to be the main mechanisms altering the visual signalling in the transgenic mice. These were responsible for the results of the analysis that revealed a unique phenotype of reduced rod sensitivity, and impaired activation and slowed recovery of the phototransduction cascade.

Never could the simple Provencal butcher, Jean Nougaret, have imagined what legacy he would leave to the inhabitants of little Vendemian, not to mention to science. Or that some mice, biochemically adapted by humans, would one day share his defective gene with his 2076-plus descendants.

THE ENGINEER'S NEW COAT

The famous 'green' locomotive livery that was actually mustard-yellow

Suppose your boss announced to you that black is white. Would you agree if that was the price of keeping your job? Or suppose it was your employee who made this bold statement. Would you humour them if you knew they were vital to the company's success? There is a 150-year-old mysterious case of 'brown is green' that still divides opinion around the world as to how it came about and how the engineer concerned got away with it.

The man at the centre of the controversy is William Stroudley (1833 – 1889), one of Britain's foremost steam locomotive engineers. He designed some of the most famous and longest-lived locomotives, some of which are still running on heritage railways. His beginnings were relatively humble, as one of three sons of William senior, a machinist in a paper mill at Sandford-on-Thames, near Oxford. His first employment was with his father at the mill but, in 1853, he was given the opportunity to train as a locomotive engineer with the Great Western Railway and later with the Great Northern Railway.

Stroudley was clearly talented and, in 1861, landed his first big appointment as manager of the Edinburgh and Glasgow Railway's Cowlairs Works. In 1865 he moved to Inverness as

Locomotive Superintendent of the Highland Railway, becoming their Chief Mechanical Engineer the following year, a post he held for four years in all. One of the changes to the locomotive stock that he introduced in his time at Inverness was the one that was to cause all the controversy. It was a new livery, one of his own devising, to be called 'Improved Engine Green'.

The problem was that Improved Engine Green (IEG) wasn't green. It has been compared with gamboge (a dark mustard yellow pigment), or a golden yellow ochre. As much argument continues as to the precise nature and description of the colour and whether modern reproductions exactly match the original, as surrounds the question of why Stroudley so named his new colour in the first place. The obvious answer to the latter conundrum is that Stroudley was colour-deficient. This may be so, but there appears to be no other anecdotal evidence of him being involved in colour-related disputes; neither have any of his medical records been unearthed that might shed some light on the matter. Did Stroudley, one wonders, ever try the Holmgren Wool colour test, when it came into use in 1876 after a Swedish railway crash?

Other theories have been advanced for Stroudley's naming of IEG, with the assumption that his colour vision was normal. One is that, to get the Board's approval for the change of livery, he described it as 'an improvement on (the existing) Engine Green'. Proof of this might be found in Board meeting minutes. However, the National Archives of Scotland hold Highland Railway minutes of meetings of shareholders, directors and committees for the period 1865 – 1870, but no references to IEG can be found therein. The NAS also has minutes of meetings of the board of directors and committees of the Edinburgh and Glasgow Railway encompassing the time Stroudley was at their Cowlairs Works, ie for 1860 – 1865; again there is no inkling of Stroudley's possible nascent thoughts on livery colours.

In an article entitled, *Could Stroudley tell Yellow from Green* (Evans, Model Engineer, 15 July 1964, pp 521-3), the author muses (assuming Stroudley to be colour-deficient), 'But it is reasonable to suppose that one of his assistants would have pointed out the true colour; from what we know of him, Stroudley was not an unapproachable man by any means.' His biographers, though, tend to contradict this reading of Stroudley's character, being generally of the opinion that he was somewhat of the typical Victorian autocrat. More probable was that no-one dared to contradict the boss out of fear for his job.

Stroudley's influence was to become more widely felt when he moved to the other end of Britain to become Locomotive Supervisor of the London, Brighton and South Coast Railway (LBSCR). He soon introduced his IEG livery here too. According to his biographer H J Campbell Cornwall, in *William Stroudley, Craftsman of Steam* (1968), the Highland connection led to an alternative name for the colour, of 'Scotch Green'. But there is yet another theory for IEG which is that because, on Stroudley's arrival at LBSCR, passenger locomotives were a dark Brunswick green, and before that had been a dark bottle green, the name IEG was intentionally sardonic: Stroudley was just fed up with endless shades of green. He did persevere with green, however, for his goods locomotives. The dark olive green, or 'Goods Green', was apparently produced by adding black to IEG. It was reputedly based on the colour of an ivy leaf given to Stroudley by his gardener.

To see why Stroudley's influence and the argument over IEG has become so widespread, one needs to look at what he achieved in his time with LBSCR. Being a great advocate of standardisation, he first introduced a new class of locomotives eventually comprising 36 engines, which successfully managed LBSCR's express traffic for many years. A prototype of this B1 class, No 214 *Gladstone* (pictured), was completed in 1882 and, by 1927, when it was withdrawn for preservation, had covered 1,346,918 miles. It is now on display at the National Railway Museum in York. In 1872 he introduced the 'Terrier' class tank engines which were largely used for London suburban services, some being still in active use into the

1960s. Stroudley's engine designs dramatically improved the performance of locomotive stock, and he was awarded the George Stephenson medal of the Institution of Civil and Mechanical Engineers for his 1884 paper, *The Construction of Locomotive Engines* (Min. Proc. Instn. Civ. Engrs., 1884/5, 81, 76, Paper No. 2027).

Stroudley's fame and the reliability of his engines led to several examples being preserved, one Terrier class engine even finding its way to the Canadian Railway Museum. It has also resulted in his engines being popular with model railway enthusiasts; and as these aficionados can be very particular about the accuracy of every detail of their models, one will find heated debates about the exact shade of proper IEG raging worldwide across the internet. It is said that the best match for IEG can be seen on a model of the locomotive *Como* in Brighton Museum, which was reportedly painted at the LBSCR's Brighton works when the colour was still in use. *Gladstone*, and the A1 Class *Boxhill* (also on display at York) are in a version of IEG. On the Bluebell Railway, there are other notable Terriers still running: *Fenchurch*, the oldest; and *Stepney*, who featured in one of Rev W Awdry's original Railway Series stories, *Stepney the 'Bluebell' Engine*, sharing an adventure with Thomas the Tank Engine and friends.

Stroudley died in 1889, after contracting a severe chill that progressed to pneumonia while conducting a test run of one of his locomotives at that year's Paris Exhibition. A further testament to his fame is the newspaper reports of a crowd numbering several thousand that attended his burial in the Woodvale cemetery in Brighton. Unless new information comes to light from as yet undiscovered archive sources, it would appear that Stroudley took the secret of IEG with him to his grave.

FROM LENSES TO EUGENICS

The lens pioneer who turned to bespoke babies

Here we revisit a protagonist of a previous story, *Forged by War*, to explore what must rank as one of the more bizarre optical connections described so far. Put simply, Robert Klark Graham went from making spectacle lenses to manufacturing babies.

We met Robert Graham previously as the Univis Lens Co sales manager who managed to negotiate the purchase of a tank of seemingly useless CR39 plastic material. At the time there was no obvious commercial use for it, and no-one even knew whether the contents of the tank were still in a usable form anyway. In *Forged By War* (page 20), we saw how he set up his own company - which later became Armorlite - and, having brought with him most of the Univis research team, realised the resin's potential as a plastic spectacle lens material. He eventually sold Armorlite to 3M in 1979 in a multimillion-dollar deal that enabled him to finance his longstanding ideas about improving the intelligence of the human race; or at least increasing the stock of intelligent humans.

Graham's philosophy of eugenics - 'intelligent selection' - as he called it, had its roots in rural Harbor Springs, Michigan, where he was born in 1906, the son of a dentist. He noticed that his father's professional friends and colleagues tended to have relatively small families, many with only one or two children, while blue-collar workers often had larger families. His post-college experience of ten years as a salesman, calling mainly on doctors, reinforced his earlier observations of professionals having few children and prompted him to write a book setting out his thoughts on the matter.

The Future of Man, which Graham self-published in 1971, argued that Mankind was heading into an evolutionary cul-de-sac; modern life and the welfare society had resulted in the survival of weaker specimens who would have been culled by the harsher conditions that prevailed in more primitive times. Not only that, but these 'retrograde humans', in his terminology, were reproducing at a faster rate than intelligent, educated people. The resulting genetic decay would lead to the degradation of human intelligence and, in all probability, global communism. To rectify the situation the prime intelligent specimens – specifically white males – must be encouraged to father more children. Graham proposed to kick-start the process by founding a sperm bank with donations from elite scientists who were also Nobel laureates, to be distributed to suitably intelligent married women.

To this end he set up a laboratory at his ranch at Escondido, near San Diego, California, in collaboration, he claimed, with the noted geneticist and Nobel Prize winner, Hermann J Muller. The facility was to be called the Repository for Germinal Choice: according to Graham the name was Muller's idea. But the equipment sat idle for several years firstly, because, claimed Graham, Muller was very wary of adverse publicity and was concerned about getting the technicalities just right and, secondly, even after Muller's death in 1967, Graham was still preoccupied with Armorlite.

The sale of Armorlite was Graham's cue to go live with the Repository. An initial difficulty was persuading Nobel science laureates to donate sperm. According to different sources either two or four agreed anonymously. Only one other, William Shockley, consented to his name being made public. Shockley, co-inventor of the transistor, for which he won the 1956 Nobel Prize for Physics, also held controversial views on eugenics; his co-operation with Graham's programme probably did it more harm than good.

Graham soon realised that the relatively advanced age of Nobel laureates was not ideal in terms of viability of sperm so he widened his net to include firstly, younger scientists, and then eminent scholars in other fields as well as high-performing athletes. But the absence of serious genetic conditions and certain minor ailments (including myopia) were insisted upon. He compiled a catalogue of donors, giving each a pseudonym and a description. Examples included 'Mr Grey-White....ruggedly handsome, outgoing, and positive, a university professor, expert marksman who enjoys the classics' and 'Mr Fuchsia,' an Olympic gold medallist, 'Tall, dark, handsome, bright, a successful businessman and author.' Other attributes were also listed, such as ethnic background, eye, hair and skin colour, height, weight and general appearance. The catalogue was mailed to women who had responded to an advertisement in a Mensa magazine for well-educated, financially comfortable women who were married to infertile men.

The Repository came to prominence – or notoriety – when the *Los Angeles Times* ran an article on it in 1980. Reaction was uniformly bad, with critics comparing Graham's institution with Nazi eugenics and deploring this apparent step towards 'designer babies'. Graham's estate was picketed. And then, not much seemed to happen for a while. Interest faded until news of the first birth facilitated by the Repository was broken by the *National Enquirer*. The success rate was actually quite low, as the process entailed a regime of self-insemination by the woman of sperm mailed to her in small vials. Nevertheless there were around 15 – 20 births per year during the late 1980s and early 1990s, and a steady waiting list of women to get on the programme.

The whole operation was funded by Graham as a not-for-profit organization, with no fee paid to donors and none asked (except for expenses) from recipients. Consequently, the Repository effectively came to an end with Graham's death in 1997 as a result of a fall in a hotel bathroom. Active until the last, he was attending a meeting of the American Association for the Advancement of Science in Seattle, the type of conference at which he would seek out young, promising scientists as potential donors.

What is the legacy of the 'Nobel Prize sperm bank'? None of the children conceived were the offspring of Nobel Prize winners. In all, the Repository claimed 229 successful births across the United States and beyond. Mostly, the identities of donors, recipients and children remain unknown; no-one knows what happened to the Repository's records when it closed down. But one baby became almost a poster child for the programme. Now aged 28, Doron Blake was put in the media spotlight almost as soon as he was born. At two years old he could use a computer, and was studying *Hamlet* and algebra in nursery. Yet he grew up to be an ordinary, socially-awkward student, albeit with a high IQ. Efforts have been made to trace donors, parents and children associated with the Repository, notably by the journalist, David Plotz, via features for the online magazine, *Slate.* He has had some success, but the majority contacted still wish to remain anonymous.

The Repository inadvertently did have one positive outcome. Until Graham's idea, artificial insemination by donor was a fairly brutal affair where women were at the mercy of their doctor's regime, forced to accept donor sperm of unknown origin. After Graham, sperm banks became customer-friendly operations allowing the potential parents to take control of the process; and choosing sperm from large catalogues supplying detailed profiles of donors becoming the norm.

Here's a final thought: there is a certain sort of curious irony when considering the juxtaposition of Robert K Graham's two careers as maker of spectacle lenses and babies. His selection criteria for the raw material to create the latter should have made it less likely for those produced to have need of the former.

AN OPTICAL TELEPHONE

The surprising identity of the 'father' of fibre-optic communications

Below is described a novel apparatus for a wireless optical telephone. The name of the Scottish-born inventor may come as a surprise since, although rather well-known, he is not generally thought of in connection with optical research. His identity will be revealed later on.

The inventor is on record as saying that be believes the invention of what he has termed the 'photophone' is his most important yet. He has even revealed that he wished to name his new daughter 'Photophone' but his wife sensibly prevailed with her choice of Marian. His father describes his son's palpable excitement over his technological breakthrough: 'When he contacted me about it, he said, "I have heard articulate speech by sunlight! I have heard a ray of the sun cough and sing! .... I have been able to hear a shadow and I have even perceived by ear the passage of a cloud across the sun's disk. You are the grandfather of the Photophone and I want to share my delight at my success."'

Is the hyperbole justified? The idea behind this novel optical technology and details of the construction of prototypes is set out in a paper in the *American Journal of Sciences* entitled 'On the Production and Reproduction of Sound by Light' and was read by the author before the American Association for the Advancement of

Science (AAAS) on 27 August, in Boston. Here he describes the first public outing he gave to the idea of using a photoelectric detector made of selenium to detect a modulated beam of light in order to reproduce audio signals. This was in a lecture given at the Royal Institution some two years previously. As a result he received communications from other investigators reporting sound effects obtained from exposing selenium, which had been wired in to an electric circuit, to light. This spurred on his own experiments, aided by his assistant, Charles Sumner Tainter.

He acknowledged, in his AAAS presentation, that, 'Although the idea of producing and reproducing sound by the action of light.......was an entirely original and independent conception of my own, I recognize the fact that the knowledge necessary for its conception has been disseminated throughout the civilized world, and that the idea may therefore have occurred, independently to many other minds.' In fact, earlier in the year, he and Tainter, using their unique equipment, had already sent the first ever communication by wireless optical telephone between buildings. The message was sent on 1 April from the roof of the Franklin School, Washington, DC, across the road to their laboratory at 1325 'L' Street, a distance of 213 metres.

That this message was transmitted by the simple medium of sunlight only serves to illustrate what potential the system may achieve with the co-option of lasers, fibre-optics and advances in photoelectric detectors. Indeed, the inventor has four patents relating to the photophone, more than a fifth of the total that he has sole name to. Not that the *New York Times*, commenting on the experiment, was impressed. In an editorial of typical journalistic ignorance of science it stated, 'The ordinary man.....will find a little difficulty in comprehending how sunbeams are to be used. Does [the] Professor.....intend to connect Boston and Cambridge.....with a line of sunbeams hung on telegraph posts, and, if so, what diameter are the sunbeams to be.....will it be necessary to insulate them against the weather.....until (the public) sees a man going through the streets with a coil of No. 12 sunbeams on his shoulder, and suspending them from pole to pole, there will be a general feeling that there is something about [the] Professor['s] photophone which places a tremendous strain on human credulity.'

Not for the first time has the reporting of science in the newspapers been found wanting. In contrast, the inventor is on record as speculating: 'Can the imagination picture what the future of this invention is to be!.....We may talk by light to any visible distance without any conduction wire.....In general science, discoveries will be made by the Photophone undreamed of just now.' Indeed he has further discussed its possible application to the spectral analysis of stars and sunspots.

The basic construction of the photophone as used for that initial wireless communication is in essence quite simple – as the best ideas usually are. A lens or lenses collects light from a source (in this case sunlight) on to a thin plane mirror placed at the end of a 'speaking tube' which forms the transmitter. The sound pressure waves from a voice speaking into the tube cause the mirror to flex in response, which in turn modulates the intensity of the light beam reflected off it. This beam is then reflected by a parabolic mirror at whose focus is mounted a photoelectric cell of crystalline selenium acting as a receiver. The cell is connected to a battery and telephone apparatus which transduces the now-electrical signal back into audio form.

The principal patent for the photophone, which describes the above in much more detail is US Patent 235199, 'Apparatus for Signalling and Communicating, Called "Photophone"'. An obvious question remains: if this technology is so truly revolutionary, has been written up in scientific papers, reported at conferences such as the AAAS, reported in the Press and patented, why isn't everybody talking about the photophone? Well, Patent 235199 was registered on 7 December 1880, to one Alexander Graham Bell. He really was too far ahead of his time with this idea. Unfortunately sunlight was prone to too much interference to be able to reproduce a goodenough quality signal, and other available light sources severely limited the range of the apparatus; although, by 1897, he had managed to increase it to several kilometres. But his device is still considered to be the forerunner of fibre-optic telecommunications.

Bell (1847 – 1922) was initially inspired to explore the mechanisms of speech and hearing by his family's expertise in teaching elocution and by his mother's deafness. Of course this led to his invention of the telephone, but his scientific legacy ranges widely. He is also credited with inventing the metal detector, the hydrofoil and an early form of iron lung. He experimented unsuccessfully with methods of imprinting a magnetic field onto a record; but the principal was sound, as proved by the tape recorder, floppy disc and hard drive. He was a pioneer of aviation research and made notes on ideas for aeroplanes, parachutes and vertical take-off and landing systems. He installed a primitive type of air-conditioning in his house and speculated on the necessity for alternative forms of energy such as solar panels and methane from waste. Bell was also a co-founder of the National Geographic Society. And he invented the photophone.

In 1929 a scientist at Bell Laboratories proposed a unit of measurement to quantify levels of sound pressure and power. He suggested naming it the 'bel' (B) in Bell's honour. As defined, the unit is too large for normal use, so the decibel (dB), or 0.1B, is generally used. Bell explained the motivation behind his career at the 1891 Patent Congress in Washington, DC: 'The inventor is a man who looks around upon the world and is not contented with things as they are. He wants to improve whatever he sees, he wants to benefit the world; he is haunted by an idea. The spirit of invention possesses him, seeking materialization.' There could not be a better summary of his life's work.

BLIND JACK, OF MANY TRADES

The extraordinary achievements of a multi-talented Yorkshireman

It would be quite some achievement for anyone to claim: to be an accomplished musician with two instruments, to help raise a militia and lead those soldiers into battle, to make money from running a transport business and from trading in legal (and not so legal) wares, and to build roads and bridges, some of which are still in use over 250 years later. How much more so if that person was blind? The *Eccentric Mirror* magazine posed the question to its readers nationwide in 1807, 'Who would expect a man, totally blind from his infancy, superintending the building of bridges and construction of highroads?'

John Metcalf, born in the Yorkshire town of Knaresborough in 1717, is who. His parents were of middling working class but not poor, so were able to pay for John's education at a small private school and for music lessons. He began school aged four but, two years later, tragedy struck when he contracted smallpox and was rendered blind as a result. John obviously possessed a determined character since within six months of recovering from the disease he was already able, without outside help, to walk from his home to the end of his street and back. After three more years, he later claimed in his autobiography, he could find his way to any part of Knaresborough; he craved the freedom of the outdoors, playing with other boys and going for five- or six-mile walks into the countryside.

It is possible that young Metcalf learned to read by feeling the tombstone inscriptions at the nearby churchyard, much as his older Yorkshire contemporary, the mathematician Nicholas Saunderson, also blinded from infancy, had done. The first embossed books were not developed until 1784; Braille in 1829. He began violin lessons at thirteen, and within a year was playing at county dances. At fifteen he accepted an invitation to become the official musician at Harrogate spa. There he gained a reputation through the quality of his fiddle-playing (he also played the oboe) and his handsome six-foot-one stature; despite the milky appearance of his smallpox-scarred corneas.

Metcalf was to use this reputation, and one of the connections he made with the gentry, to help a local worthy, William Thornton, recruit for a militia in response to the 1745 Jacobite Rebellion led by the Young Pretender, Bonnie Prince Charlie. Thornton's troop was ordered to join General Wade's forces at Newcastle, where they joined up with a regular regiment and marched to Hexham. There they fell in with the King's men who were pursuing the rebels on their retreat to Scotland; all in the depths of a snowy winter. On January 17, 1746 they saw action in the Battle of Falkirk, where Metcalf led the 'Yorkshire Blues' into battle, playing his oboe.

The Royalists suffered defeat at Falkirk but, under the Duke of Cumberland (the second son of George II) the tide would turn at Culloden. The Duke came to hear of Metcalf and, according to Metcalf's account recorded in the third-person, 'His Royal Highness took notice of Metcalf and spoke to him several times on the march, observing how well, by the sound of the drum, he was able to keep his pace.'

After returning to Knaresborough and some involvement in various nefarious forays into contraband trade, Metcalf settled on starting a wagon-transport service between Knaresborough and

York. But a previous slow, uncomfortable coach trip to London some years before that had persuaded him to make the return journey on foot (in five days), had revealed to him the appalling state of the roads. His time with General Wade, who had helped to control the Scottish Highlands after the 1715 Rebellion by constructing good roads through the countryside to enable efficient troop movements, showed him what could be done about it. When he got wind of a tender to build a section of road on one of the growing number of turnpikes springing up, he used his contacts to obtain the contract. He recruited labourers much as he did soldiers for Thornton's militia and used his natural leadership skills and military experience to oversee all aspects of the work and make the project a success. That first road built by Metcalf was the three-mile stretch between Minskip and Ferrensby, completed ahead of schedule and to a highly satisfactory standard. It is still part of the Knaresborough to Boroughbridge route, now designated the A6055.

'Blind Jack', as Metcalf had become known, was extremely good at mental arithmetic and estimating quantities. He had a 'viameter' constructed to enable him to make accurate measurements of distances. It consisted of an iron-rimmed, spoked, wooden wheel of almost three feet diameter; extending upwards from the wheel's hub on either side were two struts containing a mechanism to turn a pointer on a calibrated brass dial situated below the handle at the top, from which Metcalf could 'feel' the distance measurement. This instrument was an early version of the same device, based on the same principle, as used by modern workmen. A representation of the viameter can be seen next to Metcalf's seated statue (Fig 1); the original is housed in the Knaresborough Courthouse Museum.

After the success of his first road Metcalf diversified into bridge-building, winning a contract for a new bridge over the small River Tut that runs through Boroughbridge. With characteristic efficiency the eighteen-feet-span structure was built quickly and well and is still in use essentially in its original form. But the road-building work, including a section of the A61 between Harrogate and Harewood, expanded to the point where he received his biggest payment, of £6,400, for four and a half miles of new roads. Still aged only forty, he realised that civil engineering was destined to be his full-time career; one which lasted another thirty-six years.

After the death of his wife, Dolly, mother of his four children and who acted as his accountant, and due to his advancing age, he gradually began to wind down the business. That is not to say he became inactive; he dabbled in the wool, hay and timber trades and found time, in 1795, to dictate his memoirs to publishers in York. They printed *The Life of John Metcalf*, an autobiography transliterated from his broad North Yorkshire accent into a third-person account.

Metcalf was generally of vigorous health by all accounts, but Time overtook him and, at the age of 92, he died on April 26, 1810. He left behind four children, twenty grandchildren and 98 great- and great-great-grandchildren. A visitor once recounted Metcalf's attitude to blindness thus: 'He thought Providence knew what was best for us. His disposition was enterprising, and had his sight been spared it might have been worse for him.' The verse epitaph on Metcalf's headstone, sponsored by a local philanthropist, began:

'Here lies John Metcalf, one whose infant sight
Felt the dark pressure of an endless night;
Yet such the fervour of his dauntless mind
His limbs full strung, his spirits unconfin'd...'

One likes to imagine a young blind person wandering through All Saints, Spofforth, churchyard, taking inspiration from those words after 'reading' the inscription by touch.

My thanks to Frank Norville for suggesting this topic and supplying me with a copy of Blind Jack of Knaresborough *by Arnold Kellet, The History Press Ltd, 2008.*

CHEERS, MR WAKEFIELD

The man who connects motor oil with the world's oldest optical organisation

When you start up your car engine, spare a thought for what the well-known television advertisement calls the 'liquid engineering'. Without a lubricant sufficiently runny to work efficiently from a cold start yet viscous enough to continue to work at high temperatures, your engine is taking your car nowhere. Then give thanks for the lubricant oil developed in the early twentieth century by a man who twice led the oldest optical organisation in the world, the Worshipful Company of Spectacle Makers (WCSM).

Charles Cheers Wakefield was born in Liverpool in 1859, the son of a customs official, and obtained his unusual middle name from his mother's maiden surname. He could have followed his father by taking a safe civil service post but instead joined a local oil-broking firm. He then obtained a post at the London office of an American petroleum company, Vacuum Oil, where he was subsequently appointed manager for the British Empire. This job entailed extensive worldwide travel but, in 1888, he married Sarah, the daughter of a Liverpudlian bookkeeper, and a year later set up his own company, C C Wakefield & Co, based in London.

Initially the company dealt in lubricating oil and appliances, then the provision of lubricants for British and overseas railways. Alive to the growing popularity of the then novel forms of transport, the motor car and the aeroplane, Wakefield saw the opportunity to develop new forms of lubrication suited specifically to these new types of engines which other companies were largely ignoring. He developed Wakefield Motor Oil, having found that the addition of a quantity of castor oil, a vegetable oil derived from castor beans, produced the qualities required of the new lubricant. This vital ingredient was reflected in the lubricant's brand name: Castrol.

The aeronautical engineering challenge was to produce a lubricant that remained fluid at the temperatures of as low as -32°C that were encountered at high altitude. In 1909 Wakefield launched the Castrol R lubricant for aircraft, the same year that the first British aeronautical racing event was held, at Doncaster. He had the satisfaction of seeing that all the winning aircraft were using his product. These developments were to play an important part in the First World War, given the increased mechanisation of warfare. Even the Kaiser's chief of staff had to admit to his boss that the British formula for a non-freezing aircraft lubricant had thus far eluded the Germans. In 1917, with agriculture also becoming more mechanised in an effort to plug the gap in food supplies disrupted by the U-boats, Wakefield produced another variant of his lubricant, Agricastrol.

The success of Castrol, and a trusted management team provided by Wakefield's brother-in-law and managers who followed Wakefield from Vacuum Oil, meant that the company was being run like – excusing the pun – a well-oiled machine. With his marriage childless, Wakefield decided to throw himself into City of London politics becoming, in the process, an enthusiastic supporter of the City's work and traditions. He was elected to the Court of Common Council in 1904 and served as a sheriff in 1907-8, becoming an alderman and receiving a knighthood in the latter year. He was also admitted as a Freeman of the WCSM in 1906.

It is not clear why exactly Wakefield began an association with the WCSM, although the Company, at least until the end of the nineteenth century, had tended to be more interested and involved in City affairs than its founding craft. Wakefield's 'mother company' was actually the Haberdashers', which he served as Master; but he also served as Master of the Cordwainers' and Gardeners' Companies. That is not to say he did not devote much energy to supporting the WCSM. He held the two senior posts of Renter Warden in 1918 and Upper Warden in 1919, 1922 and 1923, and served as Master on two occasions, in 1920 and 1924-6. His efforts were clearly appreciated too, as the WCSM archives show that the Court presented him with a silver salver on April 26, 1923. Among all these activities he still found time to serve a term as Lord Mayor, in 1915.

Meanwhile, Wakefield's strategy to promote his brand was via the twin prongs of publicity and research. Advertising at aeronautical and automotive events was pursued, as well as sponsorship of individual feats of endurance and record-breaking. During the 1920s and 1930s the world land speed record was broken 23 times, with Castrol in the engine on 18 occasions – successes that were followed up with substantial newspaper advertisements applauding the achievements. In aviation, Alcock and Brown used Castrol R to help them achieve the first non-stop transatlantic flight in 1919, and Wakefield part funded Amy Johnson's 1930 solo flight to Australia. Close association with these events enabled Wakefield's scientists to obtain much useful data for advanced research at the company's Liverpool plant and later at a new facility in Hayes, Middlesex. By 1960, with Wakefield long dead, the brand was more famous than his name, so a decision was taken to rename C C Wakefield & Co as Castrol Ltd. In 1966, Castrol was bought by Burmah Oil, which itself was later swallowed up by BP, a company it once largely owned. But while the Burmah name disappeared, the Castrol brand lives on.

Wakefield became extremely wealthy through Castrol but he was generous with his time and money, and his philanthropy touched many areas of London life. He was president and benefactor of Bethlem Royal Hospital and a governor of St Thomas' and St Bartholomew's hospitals and the National Children's Home and Orphanage. The Guildhall Library and Art Gallery benefitted from his financial help, and various institutions were grateful for the donation of rare artefacts. But perhaps one of his most enduring legacies was his setting up of the Wakefield Trust, which survives today as the Wakefield and Tetley Trust.

Wakefield established his Trust in 1937 with the help of Rev Phillip 'Tubby' Clayton, a former army chaplain who had established a soldiers' rest house in Belgium during the First World War under the name 'Toc H' and had brought the idea back to England. At the time he was vicar of All Hallows by the Tower in the City. Wakefield endowed the Trust with several houses in the vicinity of Tower Hill to be used 'for such charitable purposes as will be most conducive to the development of Tower Hill and Trinity Square as a centre of welfare work or as a centre from which welfare work can be conducted.' One of the suitable purposes set out was a headquarters for Toc H. In 2006 the Trust merged with the Tetley Trust, a charity with overlapping aims, the resultant body now serving the inhabitants of the City, Tower Hamlets and Southwark.

Wakefield's extensive work in industry, for the war effort and charitable causes was recognised by the granting of several honours. In addition to his knighthood, in 1916 he was made a baronet, and a CBE followed in 1919. In 1930 and 1934 he was made successively a Baron, then 1st Viscount of Hythe. He was active up until his death in 1941. His funeral was accompanied by falling snow, and 'Tubby' Clayton remarked that the scene could be interpreted as 'each flake a "Thank You" from a London child.'

DECODING THE WORLD

'The last man who knew everything'

It was a French demolition squad that found the object 215 years ago. Fortunately, they realised the importance of the stone covered with engravings of text in three different scripts built into an ancient wall in the Egyptian village of Rashid (Rosetta). With the Napoleonic Wars turning in favour of the British, the Rosetta Stone ended up in the British Museum where it has remained on display since 1802, apart from a short excursion to Paris in 1972. It took a mind that first described astigmatism and the mechanism of accommodation of the eye, and performed experiments demonstrating the nature of light and colour vision, to reveal its secrets.

Before handing the stone over, the French made copies of the inscriptions, which were in Egyptian hieroglyphs and Greek with a hitherto unknown script sandwiched between, and distributed these copies to leading European scholars. The stone, of dark grey igneous rock speckled with feldspar and mica, had its top section containing much of the hieroglyphs missing. But it was still a golden opportunity to discover the meaning of these ancient Egyptian symbols that had fascinated and confounded people for hundreds of years. The final part of the Greek inscription suggested that the stone was one of many copies decreed to be displayed at various temples, implying also that the three passages were translations of the same edict. Hence the potential key to the hieroglyphs: the race was on.

Thomas Young (1773 – 1829) was a relatively late entrant to the fray. Absorbed in giving a series of lectures at the Royal Institution covering just about the whole of known science, then writing them up for publication and devoting time to his medical studies, Young only turned his attentions to the strange symbols of ancient Egypt in 1814, when a friend brought back some papyri from travels in Egypt for him to examine. This was a man who was a child prodigy, his fellow students at Cambridge University nicknaming him 'Phenomenon Young'. His curiosity piqued, he obtained a copy of the Rosetta stone inscriptions to study, in between attending to his medical practice.

Young's genius enabled him to make several original discoveries regarding the stone's inscriptions. Firstly, he recognised that some of the symbols in the second paragraph bore, in his own words, 'striking resemblance' to corresponding hieroglyphs. This was the first hint that the second script was not of a completely different language from the hieroglyphs but was related to it; in fact, it turns out that it is a 'demotic', or common, form of the sacred Egyptian language represented in the hieroglyphs. Secondly, he overturned accepted theory that the demotic script was purely alphabetic as opposed to the symbolic hieroglyphs. His inspired leap was to realise that it was actually a mixture of the two: the demotic employed shorthand symbols analogous to, say, +, &, or % interspersed with alphabetical characters. More insights followed, such as the decipherment of hieroglyph plural markers, numerical notations and the identification of a symbol to denote a female name.

Most important, though, was his work on cartouches, the hieroglyphs enclosed in an oval border, so named by French soldiers because of their resemblance to gun cartridges (*cartouches* in French). It was known from the translation of the Rosetta Stone's Greek inscription that the text was a decree issued at Memphis by a nationwide gathering of priests on the first anniversary of the coronation of the teenage Ptolemy V Epiphanes, king of all-Egypt on March 27, 196 BC. Previous investigators had suggested that the cartouches contained royal or religious names, and that foreign names in the cartouches might be spelt phonetically. There are six cartouches on the stone: three short and three long. From the Greek translation it was almost certain that at least one had to contain the name Ptolemy. When Young compared a long and a short cartouche that had the same symbols, plus additional symbols in the long one, he realised that these extra symbols represented honorifics such as in the Greek translation, 'Ptolemy, living for ever, beloved of Ptah.' Now he was able to match letters in the short cartouche with known letters and phonetic values, many still accepted.

Young can be considered to be the first person to read Egyptian hieroglyphs in 1500 years. He compiled an English vocabulary for 218 demotic and 200 hieroglyphic words, of which around 80 are still accepted today; he assigned phonetic values to thirteen hieroglyphs; and offered a tentative demotic alphabet. All these he published in an article on 'Egypt' in a supplement to the *Encyclopaedia Britannica* in 1818. But his suppositions weren't entirely correct. If, as Young's Westminster Abbey epitaph states, it was he who 'first penetrated the obscurity that veiled for ages the hieroglyphics of Egypt', it was the French academic, Jean-François Champillion who built on Young's discoveries to become the first person to read the hieroglyphics in full.

It is incredible to think that the work on the Rosetta Stone – an obsession, almost – was a relatively small diversion from Young's scientific investigations. His first contribution to scientific knowledge came as a result of dissecting an ox eye as a nineteen-year-old medical student. At that time, it was believed that accommodation was achieved by elongation of the eyeball, by the crystalline lens moving axially within the eye or by the lens itself changing shape. Young observed fibres in the lens that he believed to be muscular in origin, which led him to the correct conclusion of the lens changing shape, but for the wrong reason. His paper on this was read to the Royal Society, and he was elected a Fellow at the young age of twenty-one. Further experimentation, mainly on his own eyes, led to a famous paper of 1801, *On the mechanism of the eye*, in which he confirmed that his hypothesis of the lens changing shape was correct (although the mechanism involving the ciliary muscle was yet to be discovered). It also resulted in the first scientific description of astigmatism (his own), though the name was only coined around thirty years later to describe Sir George Airey's condition.

The following year Young gave another lecture, *On the theory of light and colours*. This was a theoretical work, his insight informed by the work he had done recently on the wave nature of light, including his famous double-slit experiment demonstrating diffraction and interference. He concluded that the retina contained three types of receptor, each sensitive to a different wavelength of light and that the relative stimulation of these receptors accounted for the perception of all colours. This was largely forgotten until Helmholtz rediscovered it in the 1850s and developed it into a more complete theory, only confirmed experimentally in 1959.

The word polymath is the description usually applied to Young, but even this word struggles to contain the man. We have not even mentioned his work on the elasticity of solids (Young's modulus), capillary action, surface tension and contact angle of liquids (Young-Laplace, Young-Dupré equations), medicine (haemodynamics, the Young Rule for child dosage of drugs), acoustics (the Young Temperament for tuning instruments) and languages (another *Encyclopaedia Britannica* article compared the vocabulary and grammar of 400 languages). He wrote many more articles for the *Encyclopaedia Britannica*, in several cases breaking new ground in those subjects too. Yet for all this he was a modest man: he asked that all those articles be published anonymously. Appropriately, the title of a biography by Andrew Robinson published in 2006 summed him up succinctly as *The Last Man Who Knew Everything.*

THE CALIPH'S CAPTIVE

The scholar from Basra: genius or madman?

'Dam the Nile!' That was the order of al-Hakim Bi-Amr Allah, the Caliph of Cairo, ruler of the Fatimid dynasty in Egypt during 996 – 1021. His call was answered by a Persian scholar called Ibn al-Haytham, a decision which was to lead to a revolution in the theory of vision destined to resonate through centuries of Western science.

The annual inundation of the Nile in Egypt was the source of the country's fertility, hence the river being an object of worship there since ancient times. But the flooding could be capricious and also cause extensive damage to crops. Al-Hakim was one of the first Egyptian rulers to consider the possibility of controlling the Nile's flow in order to alleviate the negative aspects of flooding and to use the waters for irrigation. The little-known figure of al-Haytham, or Alhazen as he became known in the West, came to the caliph's attention when he wrote, 'Had I been in Egypt, I could have done something to regulate the Nile so that the people could derive benefit at its ebb and flow.'

Who was this obscure man to make such a boast? Alhazen was born in Basra in 965 and made a living producing and selling Arabic translations of ancient Greek texts. From this work he gained a knowledge of Greek science and mathematics. In 1010 a messenger arrived from al-Hakim with an invitation to Egypt to discuss the Nile problem with the caliph. Alhazen was not in a position to refuse one of the most powerful men in the Muslim world. In Cairo he outlined his scheme for a dam at the village of al-Janadil near Aswan, and so impressed al-Hakim that he was immediately given all the resources he required to complete the task. A shock was in store for Alhazen: the river at that location was almost twice as wide as he had calculated, and the scheme was unworkable.

Alhazen was faced with the unenviable task of confronting al-Hakim with his failure. The caliph was an unpredictable character, labelled by some historians as the 'Mad Caliph'. He once had all the dogs in Cairo slaughtered because their barking irritated him. Surprisingly al-Hakim accepted the news with equanimity and offered Alhazen a government post instead. Despite his apparent good fortune Alhazen was still nervous about his ultimate fate and decided to feign madness in the hope of released from his obligations. But his gambit backfired as al-Hakim had him detained and placed under house arrest, a situation that was to last for ten years.

It was during this confinement that Alhazen began to meditate on the nature of vision. The eye had a special place in Islamic medicine and philosophy between the ninth and fourteenth centuries. Many specialist ophthalmological treatises appeared during this time, such as *Ten Treatises on the Eye* and *Book of Questions on the Eye* by Hunain ibn Ishaq (Johannitius). The current theory that had come down from the ancient Greeks was one of extramission: that vision was a result of rays radiating out from the eyes to touch and therefore sense objects. This was the theory expounded by Plato and expanded by Galen. Scholars such as ibn Ishaq and al-Kindi were largely influenced by Galenic ideas.

Alhazen realised this was wrong. Why, for instance, did his eyes hurt when suddenly looking at the sun from darkness? And he argued that it was impossible for rays to travel, in the instant after opening one's eyes, to distant stars. He thus concluded that light affected the eye and not vice-versa; vision must work through intromission. Light must enter the eye, travelling in straight lines from objects, and stimulate the optic nerve. This radical idea overturned more than a thousand years of accepted dogma and was the first time that the mechanics of sight had been properly described.

He then gave thought as to how, if light is independent of the eye, those rays from objects are redirected into the eye so that we can see them. Using his knowledge of Euclid, Aristotle, Ptolemy and al-Kindi, among others, and by studying how light reflected from mirrors, he worked out a theory of light that correctly described its propogation, and mathematical details of reflection and refraction. Another breakthrough was to realise that all objects reflect light, not just mirrors. He postulated primary light as being that radiated by a bright source such as the sun or a fire; secondary light being primary light reflected off other objects. The sun thus provides the primary light for us to see objects during the day. This description even led to his calculating that the brightening sky before sunrise was due to sunlight reflected by the atmosphere when the sun was up to nineteen degrees below the horizon.

Alhazen began writing down and revising his theories. These became the monumental work called *Kitab al-Manazir* (*The Book of Optics*). It was to remain unsurpassed as a source of optical knowledge for half a millennium. It came to prominence in the West in its Latin translation, and is referred to by Roger Bacon. One chapter is devoted to the structure of the eye, in which he describes ocular anatomy in detail, and correctly explains how the cornea refracts light. He even theorises that the optic nerve transmits visual sensations to the brain; no-one had ever before suggested that it was the brain rather than the eye which was responsible for creating the sensation of seeing.

All this is impressive enough. But all these discoveries were made while confined in a single room, so he had none of the sophisticated equipment or human assistance that he might otherwise have called upon for his experimentation. Instead he used thought experiments and devised ingenious, simple, repeatable experiments with which to test his hypotheses. Mostly these relied on basic objects such as bare walls, occluded windows, screens, lamps and tubes, and were designed so that they could be performed by one person to confirm the proposed theories. Only one of the experiments required the help of an assistant (where a wooden block drilled with two holes is used to let light into a room). He was the first person to go about scientific investigations in such a systematic manner and, as such, helped to establish the modern scientific method.

One day in February 1201, al-Hakim, the Mad Caliph, went for a walk in the Muqattam Hills. He never returned. The historian, al-Qifti, states that when news of al-Hakim's disappearance broke, government officials released Alhazen and restored his possessions. He continued to write copiously on science, and travel widely. The Nile was finally tamed by dam at Aswan, firstly in 1902 and, secondly, as recently as 1970 with the Aswan High Dam, nearly 800 years after Alhazen's abortive attempt.

Alhazen's importance to science is commemorated in the naming of asteroid 59239, discovered on 7 February 1999, after him. One might also care to look up at the moon one night and, there, near its eastern edge, can be found the Alhazen crater.

DOCTOR DEE'S OPTICS

Optical mystery in Elizabethan times

John Dee was one of the most remarkable and mysterious figures of the first Elizabethan age. His interests ranged widely; he thought everything was worthy of scientific investigation. He was a philosopher, mathematician, antiquarian, teacher and mechanic, who applied himself to geography, navigation, music, painting, architecture and drama. And he would have seen no contradiction in his deep study of magic and the occult. He also helped sowed the seeds for a burgeoning of the then nascent optical trade, during the century after his death.

Dee's own papers give the details of his birth, which is just as well as there is no official record of it. The relevant document is a horoscope, showing his birth as: 1527 July 13 4^h. 2′. P.M. Lat. 51°.32′. In other words, he was born at 4.02pm on July 13 1527, the location being at a latitude probably corresponding to the City of London (latitudes were not then shown on maps, but those of important places were listed in astrological tables). He was precocious academically, arriving at Cambridge University in 1542, and was elected a founder fellow of Trinity College in 1546. His thirst for knowledge took him to the continent. At Louvain, near Brussels, he furthered his mathematical education and befriended the geographer, Mercator.

One of the few Englishmen Dee associated with was the distinguished diplomat, Sir William Pickering, who bequeathed him an unusual black mirror, or 'speculum'. This object neatly illustrates the lack of distinction in those times between what we would call science and magic. It was of Aztec origin, made of a naturally occurring volcanic glass called obsidian, and brought back to Europe by a member of Cortés' expedition in the late 1520s. The Aztecs used such pieces for healing and divination, and they were associated with Tezcatlipoca, the Aztec deity of rulers, warriors and sorcerers, whose name translates as 'smoking mirror'. In 1771 the mirrorcame into the possession of Sir Horace Walpole, the antiquarian, whose hand-written note refers to a 1664 poem by Samuel Butler which calls the item 'the Devil's Looking-glass'. It was later acquired by the British Museum, where it resides today.

Dee used this mirror, as well as other crystal 'shew stones' for occult purposes but, at the same time, in his writings, he was showing an interest in optics and optical mirrors, demonstrating the properties of mirrors by the mathematics of perspective. Dee explained that the science of optics could make 'thynges, farre of, to seme nere: and nere, to seme farre of. Small thynges, to seme great: and great, to seme small. One man, to seme an Army. Or a man to be curstly affrayed of his owne shadow.'

In 1564, while in Antwerp, Dee wrote *Monas Hieroglyphica*, a highly esoteric treatise on symbolic language, in which he claims to be in possession of the most secret mysteries. Yet in his Dedication of the work to King Maximilian, the Holy Roman Emperor, he claims that the science of optics, as well as those of astronomy, mathematics, linguistics, mechanics and others will be revolutionised by this book.

While in Paris, Dee lectured on Euclid so successfully that the lecture theatres were said to be overflowing. When Henry Billingsley published the first English translation of Euclid's *Elements* in 1570, Dee published a 'Mathematical Preface' to it. This was to be the spark to ignite interest in the optical, and other science-based, crafts which were slowly developing in England at the time. The Preface aimed to promote the study of mathematics by those outside the university sphere. Dee disagreed with the traditional Oxbridge trivium syllabus of grammar, logic and rhetoric, believing that the new, scientific, quadrivium of geometry, arithmetic, astronomy and music was more relevant to the times. The Preface proved to be hugely influential among 'mechanicians', the growing class of technical craftsmen such as opticians, and was particularly popular in London. Indeed the subsequent growth of the optical trade was such that within sixty years of the publication of the Preface (ie in 1629), opticians were forming their own Livery company, the Worshipful Company of Spectacle Makers, to oversee quality control of their craft.

Dee conceived of the mechanician as one 'whose skill is, without knowledge of Mathematical demonstration, perfectly to worke and finishe any sensible work, by the Mathematical principall or derivative, demonstrated or demonstrable.' In other words, craftsmen such as 'makers of optical glasses' were, in effect, applied mathematicians even if they did not have direct knowledge of the mathematics underpinning their work, and Dee perceived the role that they would play in the advancement of knowledge.

On another level Dee considered the mathematical study of geometrical optics important because he claimed that the action of all qualities of an object is derived from light and therefore is assumed to propagate in the same way as light. He was familiar with the traditional optical theory that supposed that light propagated itself in all directions in straight lines, or rays, from a source; so he theorised that these other qualities of an object did likewise. He also related the study of optics to celestial matters; in fact it was crucial to his astrological calculations. It is important to realise that in Dee's age there was little distinction between astrology and astronomy. Even such famous contemporary astronomers as Tycho Brahe and Johannes Kepler were largely making their observations and calculations to help with their astrological investigations. As Dee notes, 'The rays of all stars are double: some are sensible or luminous, others are of more secret influence,' but, in his world, they are propagated alike according to the rules of geometrical optics.

John Dee died in 1608 or 1609. As with his birth, there are no surviving official records of his death, and his gravestone is missing. At one time, at his house in Mortlake left to him by his mother, he was a man of considerable influence, with the ear of Queen Elizabeth. He had established there perhaps the finest library in the country, and had a superb collection of scientific instruments. Dee was one of three favourites of Elizabeth whom she nicknamed 'Eyes' because of the intelligence they were able to supply her with. Each devised their own symbol, reflecting their royally-given moniker, with which they signed their letters to the Queen. Dee's was two circles followed by a seven-like character whose elongated horizontal stroke covered the circles. The symbol supposedly denoted his eyes (the circles) plus the other four - and a sixth, occult - sense (the 'seven' figure), signifying that he was the Queen's 'secret eyes'.

On returning from travels abroad, he found his library ransacked and many of his precious instruments missing. The accession of James I marked a further deterioration in his fortunes as the new monarch wanted little to do with matters supernatural, and so had no need of Dee's skills. His final years at Mortlake were lived out in poverty as a widower in the care of his daughter, Katherine, having had to sell off many of his remaining possessions to survive.

Such was the sad end of one of the important early influences on the optical profession. When we look into a mirror, we might not see and converse with angels and spirits as Dee did via his 'scryer', or medium, Edward Kelley; but we might just pause and give a nod to the man from the first Elizabeth's reign who helped optics along the path to being the vibrant profession it is in the second Elizabethan age.

THE POLITICIAN AND THE PIANO

The American, Frenchman and Englishman, and the invention of bifocals

What has a Founding Father of the United States – also a scientist, diplomat, printer and publisher – got to do with optics? And what's his optical claim got to do with a form of upright piano? The first question is easy enough: it is well-known that Benjamin Franklin (1706 - 1790) is credited as the inventor of the bifocal spectacle lens. The background to that is the basis of this article. The piano connection will be revealed in due course.

Franklin was a confirmed spectacle wearer by his thirties. His real problems began, of course, when he became presbyopic. It was a straightforward process to order reading glasses of various powers at that time. A wealthy person, such as Franklin, who realised the reading power would need increasing over the years, could purchase an entire set of lenses. This he certainly did for other people.

He sent his sister 'a Pair of every Size of Glasses from 1 to 13' from England in 1771, telling her to 'take out a Pair at a time, and hold one of the Glasses first against one Eye, and then against the other, looking on some small print. By trying and comparing at your Leisure, you may find those that are best for you, which you cannot do well in a shop, where for want of Time and Care, People often take such as strain their Eyes and hurt them.'

In this letter of instruction he showed his awareness that each eye may have different needs, which is why he did not subscribe to the usual policy of the time of selecting spectacles by one's age: 'I advise your trying each of your Eyes separately because few People's Eyes are Fellows, and almost everybody in reading or working uses one Eye principally, the other being dimmer or perhaps fitter for distant Objects.' He also advised her that once she had selected the appropriate lenses she should 'keep the higher numbers for future Use as your Eyes may grow older.'

In the early 1780s, Franklin the presbyope wrote that he could not, without his reading glasses, 'distinguish a letter of even large print' although he had no doubt suffered the problem for some time previously. It meant that he had to use 'two Pair of Spectacles which I shifted occasionally, as in travelling I sometimes read and often wanted to regard the Prospects. Finding this Change troublesome and not always sufficiently ready, I had the Glasses cut and half of each kind associated in the same Circle.' By this means,' he continues, 'as I wear my Spectacles constantly, I have only to move my Eyes up or down, as I want to see distinctly far or near, the proper Glasses being always ready.' Thus he professed himself 'happy in the invention of the Double Spectacles, which, serving for distant objects as well as near ones, make my Eyes as useful to me as ever they were.'

It is unclear when exactly Franklin had the idea for using bifocals, but a clue is contained in a note he received in 1779 while in France. It was sent by H Sykes 'Optician: Privilégié du Roi' who made glasses in Paris from 1776 to 1785. In it he states that he 'cut' (rather than 'ground') a second pair of spectacles for Franklin, mentioning that he had 'been Unfortunate, for I have broke and Spoilt three Glasses.' This comment, and the high price of 18 francs, suggests that the lenses may have been bifocals. But by 1874 he was certainly wearing the pair mentioned previously.

The first known portrait of Franklin wearing bifocals dates from this year. The American painter of the work, Charles Willson Peale, was impressed with them, as they enabled him to examine his subject and then easily switch to his canvas; a better solution than hinging one set of small lenses above the other, a current fad in London. So impressed, in fact, that he had his own pair made: 'Bought 2pr Spectacles one of 9 Inch focus & the other of 18, I cut the Glasses of both pr. and put the longest focus above and the shorter below in each frame, so that I have 2pr. of Spectacles which will serve for near or greater distance.' Fellow artists Sir Joshua Reynolds and Benjamin West became early converts. Thomas Jefferson adopted 'Dr Franklin's plan of half-glasses of different focal distances, with great advantage' to become, as President of the USA, the first head of state to wear bifocals.

Franklin may have invented bifocals, but who was it that coined that word for his 'Double Spectacles'? The neologism was the idea of civil engineer and inventor John Isaac Hawkins (1772 - 1854). He was initially known as co-inventor of an early type of mechanical pencil, but became best known for his invention of the cottage piano or pianino (or 'portable grand' as he called it).

Hawkins, an Englishman who emigrated to Philadelphia, knew both Jefferson and Peale through another of his inventions, the polygraph. This was a mechanical device which linked the writer's pen to another so that a facsimile of one's signature could be produced together with the original. Jefferson used one, and Peale obtained the American rights to it. Hawkins coined the term 'bifocal' in 1824, crediting Franklin, and even patented a design for a trifocal in 1827.

Doubts do remain as to whether Franklin was actually the first to wear bifocals or was merely a populariser of them. Certainly, as a hyperope in his seventies by the time of his purchase of the French spectacles, he would have benefitted from them many years earlier; and he refers in his correspondence to another's experiment on lenses in London some years previously.

Hawkins was happy to acknowledge Franklin as the originator of bifocals, but the reality is perhaps less than certain. But if Franklin's reputation as the inventor of these lenses ultimately derives from his writing about them, he at least covered his bases. As he himself said, 'If you would not be forgotten as soon as you are dead and rotten, either write things worth the reading, or do things worth the writing.' And he would no doubt be tickled to know that (although rarely used now) the original type of bifocal made of two half lenses glued together is still known as a 'Franklin split.'

MAVERICK OF FLEET STREET

An eighteenth century optical entrepeneur

New optical products, aggressive marketing, disputed advertising claims, optical businesses expanding while others fail: just another week in the optical news, perhaps. Were *Optician* journal around in the mid-eighteenth century, it might well have been reporting on all these topics in relation to the scientific instrument-maker and retailer Benjamin Martin.

Martin came from humble stock, born in 1704, the third of six children of a farmer with land near Guildford. Largely self-taught, he became firstly a mathematics teacher, then the proprietor of a boarding school in Chichester, West Sussex. Around 1737 he began to develop an interest in optics and optical instruments. Characteristic of Martin's personality was that to whatever technical subject he turned his mind he felt he could make improvements in. Thus experiments resulted in him developing a 'pocket compound microscope' cheaper and more portable than the fashionable 'Culpeper' model. He wrote a monograph describing his new design, including a description also of a 'universal' microscope he had devised. These and other simple optical items he advertised for sale from his home.

By the early 1740s he had swapped teaching for a life of itinerant lecturing on 'Natural and Experimental Philosophy', all the while demonstrating his apparatus and advertising his instruments for sale. While spells in Reading, Bath and Norwich were moderately successful for his lectures, the logistics of getting his optical instruments from the manufacturers to his customers was proving unsatisfactory. In 1756 he decided on moving to London to open his own shop. But to trade in the City required him to be a freeman. The relatively new discipline of instrument-making did not have its own guild (the Worshipful Company of Scientific Instrument Makers was, in fact, only established in 1955, with the support of the Spectacle Makers and Clockmakers); apparently many instrument makers became freemen of the Grocers' Company, although records show that Martin joined the Goldsmiths' Company and became a freeman of the City in February 1756.

Martin found an advantageous site in Fleet Street located just two doors away from the Royal Society's home of the time at Crane Court, so that the Society's members would inevitably pass by his premises on their way to meetings. He began a vigorous campaign of marketing innovative products almost immediately, to the consternation of the many opticians and instrument makers in the neighbourhood. One of these products was his 'visual spectacles' (sometimes referred to by modern collectors as 'Martin's Margins'). He promoted them in a pamphlet, *An Essay on Visual Glasses (Vulgarly called Spectacles)*, in which he would show 'From the principles of Optics, and the nature of the Eye, that the common Structure of those Glasses is contrary to the Rules of Art, to the Nature of Things, etc, and very prejudicial to the Eyes ... and Glasses of a New Construction proposed.'

The visual glasses were designed to overcome the many inherent faults of 'common spectacles' as Martin saw them. These were chiefly that the lenses were placed in the same plane, parallel to the eyes, causing light rays to be refracted irregularly toward the eyes; the large lens size admitted too much light to the eyes, causing irregular refraction from the lenses' periphery and also excessive, harmful light, when only a particular quantity of light is necessary for perfect and distinct vision; clear glass or the usual shades of coloured glass admitted the larger, red, particles of light which are not as refractable as the smaller, blue particles; the image through correctly coloured lenses would be more perfect than clear ones. Martin introduced three major innovative features to rectify these deficiencies:

1. The lenses were tilted inward, so that their optical axes converged on to the object of regard.
2. The lens apertures were reduced from a typical diameter of 1½ inches to one inch.
3. The lenses were tinted violet (the colour 'least hurtful to the eyes' according to Martin).

He made a shop sign bearing the image of a pair of visual spectacles, and his business prospered. Within five years of opening his shop he moved to larger premises two doors away, at 171 Fleet Street. His new neighbour, optician John Cuff, most likely was not best pleased, especially as Martin adapted a compound microscope of Cuff's design to produce an improved instrument. Martin's aggressive marketing soon sent Cuff's business to the wall.

The visual glasses were popular enough to be finding their way to America, as advertised, for instance, by John Greenhow in the *Virginia Gazette* of 11 April 1771: 'Visual spectacles, of a new construction, by Martin, the celebrated optician'. But there were early criticisms of the new glasses, which Martin had already responded to in his *Essay*. To the accusation by customers that they could see no better or that the light hurt their eyes no less with the new spectacles, Martin gave the time-honoured response that it took time to adjust to and appreciate the benefit of his appliance.

Martin, in his *Essay*, was also scathing of those opticians who, seeing the popularity of his new design, were copying his glasses, even while the optical trade was ridiculing them. 'I have only one favour to ask those worthy gentlemen,' he writes, 'and that is, that, since they have taken so much pain to deprecate my inventions, they will act consistent with themselves and not imitate them. Let them who know nothing of optics, make spectacles; and those, who profess not to use their reason, buy them; I shall always find a demand for VISUAL GLASSES.' Worse still, it seems that bootleg 'branded' examples were being circulated. Martin always marked his glasses with his initials 'B.M.' but, in price lists from 1762 onwards, Martin notes, 'N.B. The visual glasses sold by peddlers ... with the initials of my name, were never made or sold by me.'

There were advertising wars, too, most notably with the prominent optician, James Ayscough, who basically accused Martin of quackery. Martin continued to advertise but never directly answered Ayscough's criticisms. This drew Ayscough into printing ever more extensive rants, the consequence of which was merely additional free publicity for Martin.

The business was by now very successful. Joint publishing ventures with his bookseller, William Owen, helped spread knowledge of his products to an even wider audience. One huge order came from Harvard to replace instruments lost in a fire. Many of those instruments supplied during 1765 – 8 are still held by the college. By the age of 65, Martin was passing much of the everyday business to his son, Joshua, and around 1778 the business was renamed 'B. Martin & Son'. But Joshua was not in the same league as a businessman, and within four years Martin was declared bankrupt. So shocked was he by this sudden decline in fortunes, he apparently attempted suicide; at any rate, a month later Benjamin Martin was dead. He was buried in the vaults of St Dunstan's, just yards from his Fleet Street shop.

Twenty-six years' worth of price lists issued by Martin's shop and a surviving copy of the catalogue of the sale by auction of the shop's stock allows one to examine precisely the huge quantity and variety of instruments that Martin dealt in. But one can only speculate how his career would have played out under the glare of the Advertising Standards Authority, the General Optical Council, the *Optician* journal letters page...

GEORGIANA'S EYES

The strong-willed duchess who survived a life-threatening eye disease

The story of Georgiana, Duchess of Devonshire, is an eighteenth century tale of a woman whose position, beauty and charisma made her one of the foremost public figures of her time in both fashion and politics. But her private life was beset by scandalous affairs, extravagance and debt. And her personal sufferings included an excruciatingly painful and life-threatening eye disease.

Georgiana Spencer, the great, great, great, great aunt of Diana, Princess of Wales, was born at Althorp in 1757 to the future Earl and Countess Spencer. Her story was popularised by Amanda Foreman with her book *Georgiana, Duchess of Devonshire* (HarperCollins, 1998), upon which was based *The Duchess*, the 2008 film starring Keira Knightley. Georgiana was only seventeen when she married the 26-year-old William Cavendish, fifth Duke of Devonshire. She made the most of being thrust into the limelight of high society, becoming a trendsetter in contemporary fashion and taste, and using her considerable influence and political skill to negotiate and fund-raise on behalf of the Whigs. She became the first woman to take to a political platform, famously trading kisses for votes during the 1784 election campaign. With Georgiana as hostess, Chatsworth House in Derbyshire, the ducal seat, became a focus for society and Whig politics.

Yet her personal life took a disastrous turn when, after providing her husband with two daughters and a son and heir, the Duke decided that she no longer satisfied him. Even more distressing was the fact that he took Georgiana's best friend, Lady Elizabeth (Bess) Foster as his mistress, creating a love triangle that was widely talked about in society circles. In order to find some measure of comfort, Georgiana embarked on a string of affairs of her own. But they only brought her more misery, and disgrace; and her lavish lifestyle led to the embarrassment of huge debts. Then, at the age of 35, she suffered a debilitating eye disease which - although almost fatal - may have in some measure helped her to regain her reputation and social position.

One evening she went to bed with a headache. She had for some years suffered from ocular ache associated with migraine. On this occasion the pain continued for a few days, by which time her right eye had swollen to the size of an apricot. The episode, described in detail in Foreman's book, is summarised here with acknowledgement. It was certainly a painful and distressing time: her children were temporarily moved away so that they did not have to suffer their mother's screaming, and her sister wrote that 'After hearing what I did tonight I can bear anything.'

Some of the screams may have been the fault of the medics. Her physician was sent for who, in turn, summoned three top eye surgeons, most notably John Gunning, Senior Surgeon-Extraordinary to the King. One of the doctors, who was of the opinion that the eye needed to be 'flushed' through, nearly strangled her in attempting to force blood up into the head to achieve this. Georgiana's mother's correspondence contains a detailed account of the situation:

'The inflammation has been so great that the eye, the eyelids and the adjacent parts were swelled to the size of your hand doubled, and projecting forward from the face. Every attempt was made to lower this inflammation so as to prevent any ulceration, but this has been in vain. A small ulcer has formed on top of the cornea and has burst, and as far as that reaches the injury is not to be recovered. If the inflammation should increase, the ulcer form again, and again burst, it would destroy the whole substance of the eye, which would then sink ... The eyelids are still much swelled and scarred with the leeches, and the little opening between them is always filled with a thick white matter. The eye itself, to those who see it (for I cannot) is still more horrible.'

The infection was to last for several months and each visit from the doctors resulted in hours of torture. But the previous agonising birth of her daughter had given her an inner strength, and she bore it all stoically. Initially her room was darkened so that she should be unaware how badly her sight was affected. Although her health recovered reasonably well, her sight did not. A month later, the right eye could only detect shapes; and during treatment the left eye had also suffered some damage resulting in slightly blurred vision. Any light or movement was still painful to the right eye. Her physician finally announced that there was nothing more that could be done but, in order to let her down gently, the room was kept darkened for a while longer so that she would believe she was still under treatment.

Over the next few years, she became more frail, the victim of regular illnesses and infections. The vision in the left eye worsened, to the extent that Erasmus Darwin, also a doctor (and grandfather of Charles), recommended electrical therapy in order to 'galvanize' the eye. This involved delivering hundreds of powerful electric shocks per minute via crude electrodes above the temples. It did not help.

Various diagnoses have been suggested retrospectively for Georgiana's illness, such as unilateral exophthalmos of endocrine origin and orbital cellulitis. I G Schraiber, in *A Dead Disease, As Illustrated By The Illness of Georgiana, Duchess of Devonshire* (Journal Of Medical Biography, May 2002; 10(2): 105-8) also considers orbital tumour, severe sinusitis and carotid cavernous fistula in a differential diagnosis, but concludes that cavernous sinus thrombosis (CST) is the most likely cause. In arguing that the symptoms most closely correspond with the latter, he sets out the clinical course of the disease: constant headache (and vomiting), followed by proptosis and chemosis and abducens paralysis (the first neurological sign); then, ophthalmoplegia and corneal ulceration. Were the ophthalmoscope invented, papilloedema, venous congestion and haemorrhages would have been seen.

The main causative agent of CST is staphylococcus aureus, but streptococcus pneumoniae, gram-negative bacilli and, rarely, fungi, may also be implicated. The infection usually spreads from the sinuses, teeth, ears, eyes, nose or facial skin, which is understandable when one considers the anatomy of the cavernous sinuses, situated either side of the pituitary fossa and just behind the optic chiasma. In the lateral wall of each sinus run the second, third and fourth cranial nerves, and the maxilliary division of the fifth. Through the centre travels the internal carotid artery, below which is the sixth cranial nerve. And the sinus receives the venous drainage from the areas of the face supplied by the maxilliary and ophthalmic divisions of the trigeminal nerve and the veins around the ear. CST is fortunately a rare disease.

Nowadays such tests as CT head scan, MRI brain scan and sinus X-ray are available to aid diagnosis, and treatment with high-dose intravenous antibiotics keep mortality rates very low. In Georgiana's day, of course, none of this was available; indeed the cause of infection was unknown. If she did have CST, she was very lucky indeed to survive. But she was left disfigured by both the disease and her doctors' treatments.

It seems that Georgiana's personality was quite changed as she recovered from her illness and started gradually to socialise again. To quote Foreman: 'Those scars released her from her fears. ll the inhibitions about whether she was beautiful enough or whether she was up to the job left her.'

THE SUPREME EXPERIMENTER

The scientist who ushered in the modern world

Michael Faraday (1791 – 1867) must be considered the pre-eminent name in the fields of electrical and magnetic research. His importance as an experimental scientist in these subjects is reflected in the fact that he is the only scientist to have been honoured with two SI units of measurement being named after him (the Faraday and the Farad). But, almost overshadowed by his seminal work on electromagnetism, are his lesser-known yet nevertheless important contributions to experimental and theoretical optics.

Faraday's first experiments relating to light, in the 1830s, were an investigation of the transmission of electricity by sparks. It followed from work done by Charles Wheatstone in which an electric current was passed along an extremely long (half a mile) wire circuit containing three equidistant spark gaps in order to examine the nature and velocity of electricity. Using a rotating mirror he was able to see that the middle spark was retarded with respect to the other two which flashed simultaneously. It was Faraday who realised the importance of this result for a theory of electricity as being a probable link between conduction and induction. He varied Wheatstone's experimental set-up and made several discoveries.

Firstly, he replaced one of the wires connecting the electrodes in the circuit with relatively poor conductors such as water or glass and found that the retardation of the middle spark increased. He then concentrated on the effects of sparking electrodes in different gases. Here he found that 'the characters of the electric spark in different gases vary,' and considered the spark colour to be peculiar to each gas. Although he did not examine the spectra of the sparks created by different electrodes and gases, Faraday's work came to be acknowledged as the theoretical foundation of the study of spark spectra. Faraday was using light mainly as a tool for his electrical investigations and, apart from some work in 1833 on the effect of passing polarised light through an electrolyte which produced a nil result, his optical studies came to a halt until 1845.

What made Faraday return to studies involving light? In the ongoing debate about the nature of atoms, Faraday held to a 'centre of force' atomic theory in which he assigned optical, magnetic and other properties to the atmosphere of forces surrounding the centre. This, he felt, had implications for 'the theory of light and the supposed aether.' Using light to investigate the nature of matter seemed the obvious thing to do: 'Light and electricity are two great searching investigators of the molecular structure of bodies,' he declared in a letter. Faraday had also had some contact with a young William Thomson (later Lord Kelvin), who was musing on what effect a transparent dielectric (a poor conductor that can support an electrostatic field) might have on polarised light. He referred Thomson to his own negative results of 1833, but was spurred into resuming his investigations. Faraday's diary for 30 August 1845 has a heading 'Polarized light and Electrolytes' under which he proposed to recommence the study of the relationship between light and electricity.

In the first experiments he passed polarised light through various solutions being electrolysed. The lack of any change in the light confirmed his 1833 results. Next he switched to solid media such as a silicate borate of lead glass that he had made years earlier. Still a nil result. Then he looked back to experiments he had carried out some months earlier on the nature of magnetism and decided to combine this work with his current experiments on light. Now, instead of electrolytes, he decided to try passing polarised light through various transparent media positioned within lines of magnetic force.

He set up electromagnets in five different arrangements and initially used air as the medium, possibly as a control experiment. The state of polarisation of the light passed through the magnetic fields was analysed with a Nicol prism. There was no effect observed with any of the arrangements with air or through various solid transparent media. Then he tried the lead glass used in previous experiments. In one of the arrangements he found that the light passing through the glass in the magnetic field had been partly depolarised. He had discovered the magneto-optical effect, now known as the Faraday effect. In his *Diary* he wrote, 'Thus magnetic force and light were proved to have relation to each other. This fact will most likely prove exceedingly fertile and of great value in the investigation of both conditions of natural force.' The discovery was a then rare instance of an important optical discovery by experiment without recourse to the analytical mathematical theory of light that was now becoming prevalent. Herschel and others had been searching unsuccessfully for this effect more than twenty years earlier, so news of Faraday's discovery was rapidly disseminated and acclaimed.

Faraday further investigated the conditions under which this new effect could be demonstrated. He found that (as recorded in his *Diary*), 'when the *polarized ray* passes *parallel* to the lines of *magnetic induction*, or rather to the *direction of the magnetic curves*, that the glass manifests its power of affecting the ray.' By using a stronger electromagnet he was able to demonstrate his effect in other materials too. Having not observed the effect with electricity, he concluded that electricity had insufficient strength to render the phenomenon visible. He also realised that the 'magnetic force does not act on the ray of light directly (as witness non action in air etc.), but through the mediation of special matters.' In other words, magnetism affected the behaviour of polarised light by influencing the matter through which the light was passing: indeed, 'a new magnetic force or mode of action on matter.'

The Faraday effect can be defined as the rotation of the plane of polarisation of linearly polarised light passed through a medium in the direction of the lines of force of an applied magnetic field. Having provided the first evidence that light and electromagnetism are related, James Clerk Maxwell developed in mathematical form the theoretical basis for this. The Faraday effect has some important modern applications in measuring instruments. Examples include 'spintronics' which studies the polarisation of electron spin in semiconductors; measurement of optical rotator power; and remote sensing of magnetic fields, especially useful in astronomy for analysing pulsars and the magnetic properties of the sun's coronal plasma.

The supreme experimenter, Faraday was *the* man of electromagnetism. He was only one of many making optical discoveries, but his contribution to optical theory was important nonetheless. 'The world little knows how many of the thoughts and theories which have passed through the mind of the scientific investigator have been crushed in silence and secrecy by his own severe criticism and adverse examination; that in the most successful instances not a tenth of the suggestions, the hopes, the wishes, the preliminary conclusions have been realized.' So Faraday wrote. It's fair to say that he realised more than most.

Thanks to Professor Frank James, Professor of the History of Science and Head of Collections at The Royal Institution, and editor of The Correspondence of Michael Faraday, *for taking the time to point me in the direction of relevant material for this article.*

THE MIRROR MAN

Sir John Soane: the ultimate interior designer

One would expect an eminent architect to understand the interior lighting requirements of any building that he is working on. Sir John Soane (1753 – 1837), most famous for his rebuilding of the Bank of England, devised some ingenious solutions to lighting his own house with the help of a famous name in optics. He also left behind a batch of personal optical bills that illustrates both his own visual history and the nature of early nineteenth century optical trade.

Soane was born into humble beginnings at Whitchurch, Oxfordshire, the son of a bricklayer named Soan. He started out as an errand boy for the architect George Dance the Younger who, on recognising his talent, took him into his office and subsequently transferred him to that of Henry Holland, where he remained until 1776. After a two-year tour of Italy he returned to England, built several country homes to commission and made a good marriage to the daughter of a wealthy builder; at which time he added the flourish of an extra 'e' to his surname.

In 1788, on the death of Sir Robert Taylor, he was appointed architect to the Bank of England. The work he effected there, enlarging and practically rebuilding the bank's entire structure, brought him wide recognition. He was elected a Royal Academician in 1802 and became, in 1806, the Academy's Professor of

Architecture. By this time Soane had also begun works on his own properties in Lincoln's Inn Fields. He successively demolished and rebuilt three houses on the north side, starting at No 12 during 1792 – 4; then No 13 during 1808 – 9 and 1812; and finally at No 14 in 1823 – 4. His ingenious use of natural light and space can still be seen at the labyrinthine No 13, which is open to the public (with free entry) as the Sir John Soane Museum. Figure 1 shows Soane in his pomp, aged 76, painted by Sir Thomas Lawrence in 1828 (note he is holding his spectacles, a sign of wealth and learning).

There is a tantalising reference in 'Journal Two' of Soane's accounts to a payment made on 29 January 1794, of £29.8.0, for '2 convex mirrors' to 'P & J Dollond.' This would have been for the first house, at No 12. But there are numerous convex mirrors to be seen in No 13, strategically placed to give the appearance of opening up space. They are listed, described and illustrated in a catalogue of Soane's furniture published in *Furniture History* (Vol XLIV, 2009), the journal of the Furniture History Society. They are generally described as early nineteenth century convex mirrors, with no attribution.

One particularly noteworthy set is a group of four mirrors located at the corners of the dome above the Breakfast Room. These create a complex series of 'fish-eye' reflections, and were presumably for the benefit of those seated at the central table. The Library has a set of plane mirrors running along the top of the tall bookcases on opposite sides of the room. The adjoining Dining Room also has large plane mirrors surrounding portraits on opposite sides of the room. In both cases the idea is to create an illusion of space through infinite regression of reflections; this is particularly effective during candlelit dinners, a tradition with which the museum continues. Sadly, there is probably no way of knowing whether any of these mirrors are also the Dollonds' work, not least because most of the Dollond archive has been destroyed by fire.

Soane began to populate his house with an amazingly diverse collection of objects, including a sarcophagus that is still located in the basement. Although this house was his home it was designed with the intention of showing off his art and antiquities collections. In addition to the mirrors, clever siting of windows and skylights maximised the natural light flow into the house enabling the exhibits to be seen to their best advantage. He even proposed opening the house for the use of his students the day before and after his Royal Academy lectures, and arranged his collections for easy access. After quarrelling with his two sons he resolved to establish the house and its contents as a museum for the benefit of 'amateurs and students.' This end he secured in 1833 by an Act of Parliament which came into force on his death.

The bills and receipts from Soane's opticians that are held in the museum archive range in date from 1813 to 1835. At the age of 60, by the time of the first invoice, he would certainly have been presbyopic for some time (there is a drawing by Thomas Cooney that shows Soane lecturing at the Royal Academy wearing spectacles circa 1812). However by 1815 he was complaining of weak eyes, occasionally 'too weak to read.' In December 1825 he had a cataract operation, but weak eyes were still a constant problem thereafter; illustrated perhaps by the fact that he seems to have visited several opticians in his quest for satisfactory vision.

The first invoice is from a Mr Richardson, dated 30 July 1813, for various items including 'A Pr of Tortoiseshell and silver spectacles with Brazil pebbles' costing £1.16.0. On 20 April 1817 he purchased from Mr Berge ('successor to the late Mr Ramsden') 'A Pr of tortoiseshell hand spectacles' at 15s. An invoice for 24 March 1824 tells us that Soane bought several items from 'R Huntley, Working Optician, 52 High Holborn, near Brownlow Street.' These premises are now occupied by an electronics store. Intriguingly, there is another invoice from R Huntley dated March 1830, which seems to have been paid promptly – but it is for work carried out in September 1827. A long time to wait for payment!

Some of the invoices are as interesting for their letterheads as for their contents. One group illustrates the progression of ownership of the business. An invoice heading of 1830 of 'McAll' becomes in late 1831 '[Gilkerson] and John McAll, Mathematical Instrument Maker[s] and Optician[s], Postern Row, Tower Hill', the words and letters in square brackets having been crossed through. But seeing as an 1833 invoice is again headed simply 'McAll', perhaps the one just referred to above was a case of using up old stationery. The business changes hands again as, by September 1835, a receipt is headed, 'Receipt from R P Keane, late McAll.' This receipt is for '1 Pair of Pearl and gold chased spectacles' for a princely £5, a sure sign of Soane's status and success.

Finally worthy of note, considering Soane's cataract operation, is an invoice dated 10 April 1832 for '3 pairs spectacles with cataract glasses' totalling £2.14.0. These he ordered from yet another optician: 'Watkins & Hill, Optical and Mathematical Instrument Makers to His Majesty & curators of Philosophical Apparatus to the University of London, 5 Charing Cross, London.'

Despite the best efforts of all these opticians Soane obviously found visual satisfaction hard to come by, and he remained open to any suggestions for improving his situation. His diary, for instance, for June 1834, records one unorthodox remedy that he tried: 'Lotion for the eyes, Pint of Cold Spring Water, Tablespoon of White Wine Vinegar, Tablespoon of cognac brandy, applied with a sponge.' Whether it helped is not recorded, although one wonders if he might have been better off drinking it.

Thanks to Susan Palmer, Sir John Soane Museum Archivist, for her help, particularly for allowing me to examine the Soane invoices and other materials.

MASTER OF ANATOMY

Brilliant but controversial author of influential book

Antonio Scarpa was not an ophthalmologist, yet he wrote a hugely influential textbook on ophthalmology. He studied under a professor who gave his name to a type of cataract, yet he severely retarded the progress of cataract surgery throughout Europe. And despite his interest in ocular disease, he was scathing about the idea that eye surgery should exist as a distinct surgical specialty. He was famous in his time and gave his name to several anatomical features, yet was shunned and alone at the time of his death. Who was this contrary doctor and why was his influence on ophthalmology so great?

Scarpa was born into a poor family in the northern Italian town of Motta di Livenzo in 1752. His early education was heavily influenced by his uncle, a priest. And, having become as a result of this tuition an excellent Latinist, he was able to pass the entrance examinations of the University of Padua at just 15 years of age. There he came under the tutelage of the renowned anatomist Giovanni Baptista Morgagni (he of the morgagnic cataract) who became so impressed with Scarpa's Latin and anatomical skills that he appointed him as his personal assistant and secretary. Scarpa received his medical degree when still only 19 years old from Morgagni himself, only a short time before the great man's death.

With the first of some notable acts of patronage in his favour Morgagni, before his death, had eased the way for Scarpa to obtain the post of Professor of Anatomy and Clinical Surgery at Modena, a mere year later; a post he was to keep for a decade. His star rising, Scarpa published important work on the inner ear in which its membranous labyrinth was elucidated for the first time. But his work on the second tympanic membrane partly mirrored work on the subject by Galvani, who became convinced that he was being plagiarised. Perhaps not coincidentally Scarpa managed now to obtain funding from the Duke of Modena for a study tour of Europe, thus keeping him away from the worst of the controversy.

The two-year tour of France, Austria and England (where he worked with the famous surgeons John and William Hunter) was a success on its own terms. But probably the most useful outcome of the trip was the friendship he made with Alessandro Brambilla, chief surgeon to Emperor Joseph II of Austria. Soon after his return to Modena he learned that Brambilla had procured an invitation for Scarpa to take up the chair of anatomy at the University of Pavia. The city of Pavia was then under the rule of the Austrian Emperor, and its university one of Italy's oldest and most prestigious. Somehow Scarpa managed to extricate himself from Modena and his previous patron on amicable terms.

Scarpa's first lecture at Pavia, on 25 November 1783, was to prove a watershed in medical education. Instead of a lecture in the classic style, Scarpa gave an anatomical demonstration showing the relationship between structures and organs while paying due attention to the physiology of those structures. The students were required to repeat his exercises in dissection so as to gain knowledge by experience, a standard technique in medical schools today.

Emperor Joseph II's patronage was important in Pavia becoming the world's premier centre of anatomical study. Scarpa's innovative teaching and research were pivotal too and resulted in him being given an additional post after four years there, the chair of clinical surgery. At Scarpa's suggestion, the emperor decreed that all the bodies of the deceased from the state hospital were to be transferred to the medical school; this at a time when Burke and Hare were still active in the cadaver 'trade' in Britain. Also at Scarpa's prompting, the medical students were equipped by the emperor with microscopes, instruments and other provisions. These bounties along with new buildings to work in attracted over 2000 students from throughout Europe to Pavia's medical school, most of whom attended some of Scarpa's lectures. Thus was his fame and authority ensured, as his students disseminated his name and works across the continent.

One of these works was his *Saggio di osservazioni e d'esperienze sulle principali malattie degli occhi* (*Treatise on Principal Diseases of the Eye*), published in 1801. As the title implies, this was not an exhaustive textbook on ocular disease; rather, as he explains in his preface, it was an account of 'the principal affections of the organ, which I have sedulously and repeatedly attended to...' The root of his interest in eyes was his early association with Morgagni, but he was not an eye specialist as such. Although ophthalmology as a distinct medical specialty did not really exist at that time he abhorred the idea of a surgeon narrowing his expertise to one field. So his preface also states, 'Professed oculists who have entirely devoted themselves to this department, and from whom great and important improvements might justly have been expected, have only contributed new theories, which, for the most part, been disproved by a minute anatomical investigation of the eye, or have merely furnished histories of cures little less than miraculous.'

The book was a staggering success. Over the next 36 years it ran to five Italian editions, two each in English and German and one each in Dutch and Spanish. Its popularity was partly due to having a star name as author. But the quality of the illustrations was outstanding too. Scarpa was also a gifted artist who produced all his own drawings for the book, as well as supervising their reproduction as engravings. He also controlled the overall appearance of the book by specifying the paper, typeface and bindings to be used. That he was able to do this for all his works speaks again of his skill in raising funds via patrons in the nobility.

Scarpa's opinions were so influential that his advocacy of traditional 'couching' over lens extraction for treatment of cataract held sway for around the next 50 years, as can be seen from textbooks on eye surgery during the first half of the 18th century. It was towards the end of that time that ophthalmology as a separate specialty began to be recognised.

His success as surgeon and author allowed Scarpa to live the good life. He had built for himself a villa outside the city and covered its walls with an impressive collection of paintings. Another fine collection, one of anatomical specimens, was lost during the many battles for the city between Napoleonic France and Austria. Invading Cossack troops emptied the containers of the specimens and drank the preserving spirits. Scarpa otherwise used the wars to his advantage by expediently aligning his allegiance to the changing political rulers to increase his own power within the university. By 1813 he was Rector and virtual dictator of the medical faculty. But his ruthlessness towards his colleagues came at a price: by 1832 he was friendless and blind. Shunned by his fellow professionals, he died that year during an attack of inflammation of the bladder.

Scarpa was respected for his work but detested as a man. On his funeral day his marble bust was defaced with the lines, '*Scarpa is dead; And I should care. He lived like a hog, And died like a dog.*' His assistant later exhumed his body and performed a detailed dissection of it; a fitting postscript to his life, perhaps. It is a final paradox that Scarpa's hugely influential book really marked the end of an era and that a corpus of modern ophthalmological literature began to appear just after the last edition of the *Treatise* was published.

MISSION: OPTOMETRIST

The race to fix the Hubble space Telescope

Space exploration, aeronautics research, astrophysics: NASA is an acknowledged expert in these and other scientific endeavours. But optometry? By its own account NASA likened itself to an optometrist when faced with the mother of all awkward 'non-tolerance' cases. The patient was the Hubble Space Telescope (HST), which had been given the wrong prescription and was orbiting 569km above the Earth at a velocity of 28,000km/h.

The 13.2m long, 4.2m wide HST, weighing over 11,000kg, was launched on 24 April 1990 by space shuttle *Discovery* and deployed a day later. The images HST started sending back, while better than anything a ground-based telescope could produce, were of a lower quality than had been expected: a diagnosis of spherical aberration was quickly made. This was a crushing blow for a project that was first sketched out by NASA in 1969 and embarked upon in earnest in conjunction with the European Space Agency in 1975. Astronaut training for the launch mission, using mock-ups in deep-water tanks, had begun as far back as 1979. So how did such a carefully-planned project go wrong, and how was it to be fixed?

The telescope is a reflector of the Ritchey-Chretien Cassegrain type, which consists of a large, concave, primary mirror that collects light and reflects it to a smaller, convex, secondary mirror. From here the light is reflected back through an aperture in the centre of the primary mirror and focused. HST is designed to house several instruments that can be moved individually into the focal plane of the mirror system. Investigations found the problem to lie with the 2.4m diameter primary mirror, ultimately caused by faulty assembly of the device used to measure its curvature.

This mirror needed to have its reflecting surface ground with a curvature accurate to within 0.032 microns (μm); were the mirror the diameter of the Earth, the curve could deviate from true by only 15cm. Made of ultra-low expansion glass, and kept at a constant temperature of about 21°C to avoid warping, its surface is coated with a 0.076μm layer of pure aluminium covered by a 0.025μm layer of highly UV-reflective magnesium flouoride. The fault in the mirror was determined to be an excess flattening of the peripheral curve of 2.2μm. This may only be one-fiftieth the thickness of a human hair, but in terms of the accuracy required it was devastating.

By retracing the steps of manufacture of the primary mirror, it was found that the contractor had relied on just a single test to confirm the accuracy of curvature (a mistake NASA learned the hard way not to repeat). This was a cylindrical instrument comprising two mirrors and a lens called a 'null corrector'; through which a laser beam is passed and reflected back from the mirror under test to produce an interferogram – a pattern of black and white lines which allows analysis of the mirror's curve. Unfortunately one of the null corrector's optical elements had been positioned incorrectly by 1.3mm, which then guided technicians to grind the peripheral curve erroneously. It was fortunate indeed that the contractor's null corrector had sat untouched in their plant for 10 years so that, together with the data from HSTs fuzzy images, the problem could be traced.

Where did NASA go from here? In its own words, 'NASA approached the correction of Hubble's nearsightedness as would an optometrist. The agency first diagnosed the telescope's vision problem, determined a prescription to fix the ailment and monitored the development of corrective optics to make sure the telescope's sight would be restored to the fullest extent possible.' So the diagnosis had been made. NASA now established a committee, the Hubble Independent Optical Review Panel (HIORP) to work out the prescription.

Using all available data HIORP came up with a 'conic constant' that described accurately the mirror's aberrant shape. New astronomical instruments being developed for HST could then have their optics corrected by, in effect, the inverse of that conic constant to make them compatible with the mirror. But for the three instruments already on board HST, a different solution would be required. A plan was devised to insert a series of coin-sized mirrors, designed to the inverse conic constant, behind the primary mirror on HST. The engineering problems with achieving this in practice were complex, but eventually an assembly of mirrors on mechanical arms which would be unfolded when *in situ*, the size of a telephone box, was developed. And so the Corrective Optics Space Telescope Axial Replacement (COSTAR) was born.

In light of the testing debacle that led to the original problem NASA concurrently set about establishing committees and design teams to devise several pieces of testing and verification equipment, which in turn were to be subjected to their own inspection regimes in order to ensure they were properly set up. The development of, and analysis by, these instruments took the best part of two years.

Finally COSTAR was ready to be deployed. It was carried by space shuttle *Endeavour* on Servicing Mission 1, launched on 2 December 1993. Along with COSTAR there was a replacement Wide Field Planetary Camera, with built-in correction for HST's defect, along with new solar arrays, gyroscopes and other equipment. Servicing Mission 3B, launched on 1 March 2002, removed the last of the three instruments that COSTAR had been designed to keep in focus and marked the end of its useful life. So, as part of Servicing Mission 4 (actually the fifth servicing mission, as Servicing Mission 3 was split into two parts), lasting from 11 - 24 May 2009, it was finally returned to Earth on board the shuttle *Atlantis*. It now resides in the Smithsonian's National Air and Space Museum in Washington DC where it is on permanent public display.

Successful as HST has turned out to be, its replacement is already under construction. In fact manufacture of the Webb Space Telescope (WST) began in 2004 and is expected to be launched, after many delays and cost overruns in 2018. This tennis court- sized instrument is an infra-red (IR) telescope designed to detect cooler, more distant objects than HST (which uses visible light) and to be able penetrate interstellar dust clouds. It will have an 18-segment beryllium parabolic primary mirror 6.5m at its widest part, with a focal length of 131.4m and an optical resolution of 0.07 arc seconds. It will be coated with 24 carat gold, increasing its reflectivity of IR from 85% to 98% and will have an operating temperature of 40K (-233.2C).

If WST is half as successful as HST, it will have done its job. Among other things, HST has allowed astronomers to set the age of the universe at 13.7 years with a high degree of certainty; has detected distant light from the universe when aged just 600 million years; has proved the existence of massive black holes; and has proved the existence of dark energy. WST will have a special orbit beyond the moon, 1.5 million km from Earth, at the Second Lagrange Point, a location that enables a stable distant orbit using the Earth's and the Sun's gravity in harness.

There will be no visits possible by the now-decommissioned shuttles, or any other manned craft given the distance involved. Returning to the analogy of NASA as optometrist, it had better take extra care to get its refraction spot-on first time; and when the 'job' comes back from the lab, it should be mighty careful in checking the work before sending the patient out into deep space.

STAMP OF GRATITUDE

Philatelic recognition of a medical pioneer whose invention has helped millions regain sight

On 16 September 2010 Royal Mail issued a new set of stamps entitled 'Medical Breakthroughs'. It celebrates major turning points in 20th-century British medicine. One of the six stamps (67p) commemorates Sir Harold Ridley's invention of the intra-ocular lens (IOL) and the surgery to implant it, thus effecting the first-ever total cure for cataract.

Ridley is in good company. The subjects for the stamps were whittled down to the final six, covering different areas of medicine, from a shortlist of 20 drawn up by the medical historian Professor Dorothy Porter. The other five stamps depict Sir James Black's synthesis in 1962 of artificial beta-blockers (standard first class rate); Sir Alexander Fleming's discovery of penicillin in 1928 (58p); Sir John Charnley's development of the hip-replacement operation, first performed in 1962 (60p); Sir Ronald Ross's discovery of the malarial parasite in, and its transmission by, mosquitoes in 1897 (88p); and Sir Godfrey Hounsfield's research that led to the invention of the CT scanner in 1971 (97p).

Ridley also joins a select few ophthalmologists to have been illustrated on postage stamps. These include Donders (Netherlands, 1935), Purkinje (Czechoslovakia, 1937) and Helmholtz (East Germany, 1950). To see why Ridley deserves to be in such exalted company it is worth looking again at the story of the man and his visionary invention.

29 November 1949. That is the momentous date, the day the world's first ever operation to implant an IOL was performed, by Harold Ridley at St Thomas' Hospital in London. Those who have seen the commemorative plaque in the hospital will have noted a different date given – 8 February 1950. There is some confusion in the hospital records as to whether the IOL was implanted at the same time or not as the cataract extraction. The senior nurse attending, who was entrusted with the crucial role of illuminating the eye with a hand-held torch insists that it was. In any case, the earlier date certainly marks the start of the whole procedure. It was the culmination of an idea that Ridley had been considering since the 1930s.

The story often told about the genesis of the idea for an IOL involves a 'eureka' moment for Ridley, after he observed that plastic shards from shattered aeroplane canopies embedded in World War II fighter pilots' eyes remained inert and no infections resulted. In fact the observation was one, albeit crucial, step on a path that Ridley had been exploring in thought and in conversation with colleagues for several years. By 1940 he, along with much of St Thomas', had been evacuated to locations across southern England. The region where Ridley worked was near RAF Tangmere, an important base close to the Sussex coast, whose squadrons were heavily involved in the Battle of Britain. On 15 August 1940, Flight Lt Gordon 'Mouse' Cleaver of 601, County of London, Squadron was shot down in combat over Winchester. His Hurricane's canopy was shattered and his eyes were filled with perspex splinters, but he managed to return to Tangmere. He was completely blinded in the right eye, but some vision was retained in the left.

Following initial examination by military surgeons he was sent to Moorfields. In all he had 18 operations on his eye and face some of which Ridley probably performed. During years of follow-up Cleaver in effect served as Ridley's pre-clinical trial for a material suited to making an IOL. An employee of Rayner & Keeler, who manufactured the first IOL, wrote in 1949 that, 'Everyone is aware by now of Mr Harold Ridley's recognition that this material [perspex] in the eye of service personnel appeared to cause no inflammatory condition, and in most cases, could be left alone as harmless intraocular foreign bodies.' ICI produced a pure form of their PMMA material used in aircraft cockpits which they called Perspex CQ (clinical quality) for those first IOLs. The 'Ridley IOL' was a biconvex lens of 8.35mm total diameter including a peripheral ridge designed to help grip the IOL as it was inserted into the lens capsule.

Unfortunately for Ridley, he had to announce the results of his early implant operations sooner than he would have wished after news of his procedure leaked out when a patient mistakenly made a postoperative appointment with an ophthalmologist namesake. Having published articles to establish priority, recounting the procedure and detailed follow-up, Ridley decided to make his first major presentation at the prestigious Oxford Ophthalmological Congress in July 1951. He was optimistic about the response he might receive, especially as the two subjects he took along with him enjoyed vision of 6/6 or better 18 months postoperatively. Instead it was to mark the start of a 30-year period of setbacks and depression for Ridley in the face of implacable hostility from influential figures, notably Sir Stewart Duke-Elder, when the idea and practice of IOLs fell almost into disuse. The idea of inserting a foreign body into the eye was before its time.

A few people kept the flame alive during this time. A protégé of Ridley's, Peter Choyce, provided invaluable support and research at some cost to his own advancement. Abroad, Fyodorov in Russia, Epstein in South and Binkhorst in the Netherlands were notable supporters. There was also pioneering work being carried out in the USA. There were many complications with IOLs, however, but mostly these were due to poor design and manufacture and, in some instances surgical technique. As recently as the mid-1980s opposition to IOLs was strong, and Ridley's health and career had been severely affected by the negativity surrounding his invention. But then a landmark paper was published in *Ophthalmology* (Vol 91, pp 403-419, 1984) entitled *Anterior Segment complications and Neovascular Glaucoma Following Implantation of a Posterior Chamber Inraocular Lens* by Apple *et al.* The main recommendation of the authors was that the entire IOL should be placed within the capsular bag – just as Ridley had advocated. Complications were subsequently vastly reduced in number, and the attitude to IOLs rapidly began to change.

This seminal paper was the result of a removed IOL being sent to Dr Apple, an ophthalmologist and pathologist, for analysis at his laboratory in the University of Utah's ophthalmology department after complications had arisen in the subject eye. A follow-up article, *Complications of IOLs: a Historical and Histopathological Review* (*Survey of Ophthalmology*, Vol 29, pp 1-54, 1984) more fully catalogued the various IOLs and their complications. It became one of the most reprinted articles in ophthalmology literature and as a result interest in IOLs exploded. When these papers were brought to Ridley's attention, he realised that his idea was finally about to achieve mainstream acceptance. The IOL has now restored vision to over 200 million people worldwide, and approximately 10 million IOLs are implanted annually.

There are a couple of interesting postscripts to the story. 'Mouse' Cleaver, the subject of Ridley's pre-clinical trial, had suffered a traumatic cataract in his remaining eye as a result of his original injuries. By the mid-1980s, after long-term follow-up, it was felt that he was ready to have the cataract removed and receive an implant. His sight was restored thanks to research on his own eye 40 years earlier, with a device of similar material to that which caused his injuries. Ridley himself finally gained public recognition, with election to fellowship of the Royal Society, and a knighthood. Sir Harold Ridley, FRCS, FRS (1906 – 2001) also latterly received the benefit of his own invention.

THROUGH A GLASS DARKLY

An inventor who benefitted spectacle wearers, cooks and mobile phone users

Question: How many pairs of Stookey lenses have you worn in spectacles? Being glass, the answer is probably not many lately, as most lenses are plastic these days; but for the principle of photochromic lenses we have the pioneering glass scientist, Stanley Donald Stookey, who died in November 2014 aged 99, to thank. And if you're reading this on your tablet or mobile – well, you have Stookey (and some frozen chickens) to thank again.

Born in the small village of Hay Springs, Nebraska, in 1915, to two teachers, he took a master's degree in Chemistry before gaining a doctorate in Physical Chemistry in 1940 from Massachusetts Institute of Technology. He had two job offers, from Corning Glass Works and Nabisco, the baking and food company. Not wanting to go into baking, he opted for glass research which, ironically, involved baking glass at high temperatures.

A fascination with the process of 'nucleation' ensued, as a way of producing crystalline structures within glass by the addition of tiny quantities of elements or compounds to the molten material, that would act as nuclei for the formation of crystals as the glass cooled. The amount and rate of cooling could control the crystallisation process. By this method he found he could add copper or gold to produce ruby glass; and sodium fluoride to make milky-white opal glass. Stookey initially concentrated on the latter, investigating its little-known chemistry in the hope that some useful applications could be found for it. By 1950 he had found a way to make it photosensitive, and registered the first of his sixty US patents for a form of this 'Fotoform' glass that, in daylight, resembles marble. A testament to the invention is the north face of the UN headquarters in New York which is clad in this material.

The ruby glass, too, could be made photosensitive, in this case to ultraviolet light, providing a method whereby photographs could be 'implanted' in the glass through its exposure to UV. Applications included an idea for copper-infused ruby glass coins to be produced with Abraham Lincoln's portrait suspended within, as a way of reducing the reliance on scarce copper resources in wartime America. The US Treasury seriously considered the possibility, but decided it was too expensive; as also proved the case for the spies' spectacles that would reveal secret messages on exposure to the right kind of light.

An accident with some Fotoform glass produced a serendipitous interlude in Stookey's photosensitive researches, resulting in his discovery of a completely novel class of material and a huge windfall for Corning. The incident occurred in 1952, when Stookey had placed a sample of his photosensitive glass in a furnace, to be heated to 600°C. On returning some time later he noticed that the temperature gauge had become stuck at 900°C. Expecting to find the interior of the furnace ruined by a molten mess, he was instead surprised to find that his sample was still intact. In retrieving it with a pair of tongs, he dropped the sample and was stunned when, rather than shattering, it bounced. 'It sounded like a piece of steel bouncing,' Stookey recalled, 'so I figured something different must have happened.' Indeed it had: Stookey had inadvertently created the first synthetic glass-ceramic.

The lithium silicate that had been transformed accidentally into a new milky-white material proved to be lighter than aluminium but harder than high-carbon steel and immensely stronger than ordinary soda-lime glass. Corning realised that this made it perfect for cooking with, patenting it in 1960 as Pyroceram and marketing it as CorningWare. The white dishes decorated with blue cornflowers became ubiquitous in American kitchens, earning Corning a fortune. But there were also other applications. Pyroceram's strength and lightness made it desirable for use in chemistry laboratories and microwave ovens, while NASA used the same manufacturing process for its shuttles' glass-ceramic nuts and bolts; and transparency to radar made it the ideal material for missile nose-cones.

Meanwhile Stookey turned his attention to the properties and possible applications of reversible photochromatic glass, ie materials that change colour when exposed to light but return to their original transparency when that light is removed. Together with his colleague, William H Armistead, he patented such a material in 1962 (US Patent 3,208,860: *Phototropic material and article made therefrom*).

In a paper published in *Science* ('Photochromic Silicate Glasses Sensitized by Silver Halides', W H Armistead and S D Stookey, Vol 144, No 3615, pp 150-154, 10 April 1964), they pointed out that 'hundreds of photochromic organic and inorganic substances' have been described. But for them to have useful applications they need to exhibit an important property, hitherto not found in any existing material: 'True reversibility of the colour change – that is, freedom from fatigue with repeated light-and-dark cycling.'

Armistead and Stookey achieved their aim by adding quantities of silver halide crystals to borosilicate glass. In essence, exposure to light would cause a chemical reaction to occur similar to that in traditional photographic film; except that this reaction is reversible. Only tiny amounts of these crystals were required: less than 0.1 per cent by volume, comprising crystals less than 0.1 microns across (about a hundred times thinner than a human hair).

Corning first marketed a glass photochromic lens using this technology in 1968, under the Photogray brand. Photobrown, Photogray Extra and Photobrown Extra followed. In the UK, Pilkington introduced their grey and brown Reactolite and Reactolite Rapide versions in the 1970s.

The success of CorningWare had helped to fund another area of research, named Project Muscle, which was aimed at developing other ways of strengthening glass. It was found that adding aluminium oxide to a glass mixture before it was submerged in a reinforcing hot potassium salt bath resulted in a remarkably strong, durable glass. It could be bent and twisted extraordinarily before fracturing and could withstand 100,000 pounds of pressure per square inch (compared with 7000psi for standard glass). The Corning scientists tested the material's limits by dropping samples from the top of their nine-storey building and by bombarding it with the aforementioned frozen chickens. Although it was marketed in 1962 as 'Chemcor' and was used in a small way for spectacle lenses and some other applications, it was really a material ahead of its time.

The advent of mobile phones, in particular Motorola's flip-top Razr V3 model of 2005 that featured a glass, rather than a plastic, screen, got Corning thinking about their old Chemcor samples. Then, in 2007, Apple came calling. Their requirement was for a huge amount of then non-existent chemically toughened - and optically excellent - 1.3mm thick glass for their new iPhone. But the Chemcor samples were 4mm thick. A huge effort to reformulate Chemcor and devise a manufacturing process for the new material resulted in the production of the first 'Gorilla Glass' by May of that year. Most devices you touch or swipe now are likely to involve interaction with Gorilla, probably in its 20 per cent stronger second or, even better third, version.

Stookey, who was still consulting at Corning aged 97, received the National Medal of Technology in 1987 from President Reagan for his research into glass ceramics, photosensitive glass and photochromic glass. In 2010 he was inducted into the National Inventors' Hall of Fame. Stookey recalled that when he started work at Corning, 'glass chemistry research had barely started. My main objective was to be a pioneer, discover new things, produce things that had never been seen before.' Or, indeed, seen through a glass darkly.

STARS IN THEIR EYES

Could the effect of cosmic rays on vision in space limit manned space travel?

'My God, it's full of stars!' So reports the startled spaceman, Dave Bowman, when the monolith on a Jovian moon opens to reveal a star gate in the novel and film, *2001: A Space Odyssey*. But astronauts at least as far back as the Apollo missions have regularly reported a real, if slightly less fantastical, visual phenomenon: strange flashes and streaks of light that are experienced even with the eyes closed, of a type that is not normally detectable on Earth.

Buzz Aldrin recalled seeing such entoptic phenomena, or phosphenes, on the 1969 Apollo 11 mission, the first to land on the Moon. The crew of Apollo 14, when 100,000 miles into their 1971 lunar journey, also noted these flashes of light which were especially evident during scheduled sleep periods and in the darker recesses of the spacecraft. NASA, suspecting that the source of the phosphene phenomenon was probably cosmic radiation, developed the Biostack experiment that was carried on Apollo missions 16 and 17. The apparatus comprised specimens of bacteria, crustaceans and insects at different stages of development sandwiched between specially prepared photographic plates to both record the radiation and its effect on the biological subjects.

Cosmic ray interaction was actually anticipated as far back as 1951 when Paul Tobias, a research scientist at the University of

California, Berkeley, became the first human subject in such research by allowing a beam of near light-speed protons (a common variety of cosmic rays) produced by their particle accelerator to pass through his head. When this produced the sensation of multiple brilliant flashes of light he predicted that cosmic radiation, especially that of high-energy galactic origin (eg from revolving neutron stars and supernovae) that was very difficult to shield against, would be a major risk factor in future space travel.

The Earth's magnetic field provides protection from much harmful solar and interstellar radiation, but a good deal of high-energy cosmic radiation still reaches the lower atmosphere. The importance of the effect of this radiation on low cloud formation and thus climate is only now being realised. Increased cloud cover results in a higher albedo, or amount of reflected radiation and, therefore, has a cooling effect on the Earth. Recent research, especially by the Danish physicist, Henrick Svensmark, has shown how long-term temperature trends correlate well with the cosmic radiation flux which in turn is influenced by solar activity (increased solar activity increases the sun's magnetic field, deflecting more cosmic radiation from the Earth). The results from Svensmark's team's CLOUD experiment, using the Proton Synchrotron at CERN, have provided compelling evidence for a cosmic ray cloud formation mechanism.

But there is a mysterious weakening of the Earth's magnetic field located roughly over the eastern Brazilian coastline. Within this so-called South Atlantic Anomaly (SAA) the 400-mile high magnetic field is reduced to a mere 100 miles above the Earth; so spacecraft in high-altitude orbits will pass through the SAA several times per day. In 1973 Bill Pogue, commander of Skylab 3, attempted to tap his microphone to create an aural record of light flashes but at times was unable to keep up with their frequency. After a six-month tour as cosmonaut on the Russian spacecraft MIR in 1996, Jerry Linenger reported that sleep was virtually impossible during transit of the SAA. Relocating to where his head lay behind lead storage batteries or the thickest walls had little effect on the light flashes.

Although it is clear that cosmic radiation is responsible for these phosphenes, it is still uncertain which particles and what mechanisms are involved. The phenomenon is a significant one, though: a 2003 survey of 59 astronauts found that 80 per cent experienced phosphenes, and 20 per cent agreed that their sleep was disturbed because of them. They were mainly white in colour, but other colours, notably yellow (ten per cent) were reported. They were mainly shaped like streaks, with associated sensations of in/out or left/right (but not up/down) motion. Some, around eight per cent, had a blob-like shape.

Some researchers posited mechanisms involving fluorescence of the crystalline lens, or a phenomenon called Cerenkov radiation occurring in the vitreous. Cerenkov radiation is observed when charged particles pass through a medium at faster than the speed of light for that medium, which is known as the medium's phase velocity (but note that the particles' velocity cannot be faster than the speed of light in a vacuum). The charged particles excite molecules in the medium which then emit photons of light as they return to their normal state. An example is the blue glow of the water surrounding a nuclear reactor.

In 1972, Budinger, Lyman and Tobias (Visual perception of accelerated nitrogen nuclei interacting with the human retina, *Nature, 10/1972;239(5369):209-11*) demonstrated that ions of sufficient energy stopping in or crossing the retina could result in light streaks. As we have seen above (a different) Tobias had already suggested that direct stimulation of brain tissue by near light-speed protons might produce phosphenes. Studies on MIR suggested that heavy element nuclei could also stimulate responses. A project, still running today, was begun on the International Space Station (ISS) to investigate the radiation environment within a space vessel. ALTEA (Anomalous Long Term Effects on Astronauts) is a multidisciplinary programme with a visual module operating in two distinct modes: an unmanned Dosimetry (DOSI) mode which continuously monitors the radiation flux in the ISS; and the manned Central Nervous System Monitoring (CNSM) mode which records subjective and objective cosmic ray interactions with an astronaut in a 90-minute session, the time taken for one ISS Earth orbit.

The ALTEA apparatus consists of a silicon detector system (SDS) comprising six Silicon Detector Units that altogether make up an array of 32 electrodes arranged in a soft helmet. Additional electrodes in the helmet record elecroretinogram, visual evoked potential and electroencephalogram readings which can be correlated with the SDS's reconstruction of each cosmic ray event, its trajectory and particle type. The helmet-wearer signals each perceived light flash by pressing a button. Seven sessions on three subjects were performed between 2006-7. Data was gathered regarding the range of elemental particles encountered and differences in cosmic ray flux at high and low altitudes. As expected there were peaks in flux through the SAA. It would appear that the majority of light flashes result from a direct interaction of an ion with the retina, although there is indirect evidence that some flashes are produced by interaction of particles with brain tissue.

Cosmic rays are clearly an important concern when considering the future of space exploration, especially with respect to longer voyages such as a mission to Mars. The possible impact of long-term interactions of these high-energy particles with the central nervous system has to be investigated further, while there is also the question of how particle collisions with molecules in the body might produce a cascade of further particles and collisions, resulting in tissue damage. The rate of unwanted cell mutations due to cosmic radiation requires further analysis too.

From 2010-12 the ALTEA apparatus was moved to a different module of the ISS and reconfigured to evaluate the efficacy of a variety of shielding materials against the incident cosmic ray flux. Certainly some form of effective protection from this potentially harmful radiation must be found if space travel is going to progress. From a visual and psychological point of view one can only wonder about the effects of lack of sleep, due to constant bombardment of light flashes, on an astronaut undertaking a two-year Mars voyage.

THE WAX WOMAN OF BOLOGNA

A female anatomist's exquisite wax models

Anna Morandi was born into a relatively humble background in the parish of San Martino Maggiore, in the centre of Bologna, on January 21, 1714. Very little is known about her early life and education so it is remarkable that she rose to become one of the foremost anatomists and anatomical modellers in the male-dominated world of her city's august university. Specialising in the anatomy and function of the eye, particularly the extra-ocular muscles, her microscopical investigations enabled her to discover ocular components never before described.

The University of Bologna can trace its origins back to 1088, making it the probably the oldest university in the world. Its medical school was founded around 1288; anatomy, especially the demonstrations held in the Anatomical Theatre, attracted students from across the western world. The sixteenth century was perhaps the acme of its fame but by the end of the seventeenth century the university was in crisis, its declining roll resulting in a significant loss of lucrative foreign student fees. To address the problem it was decided to create an Institute of Sciences. Founded in 1711, it aimed to renew the university's focus on observation and experiment as opposed to abstract theory. However the anatomical collection of mummified human cadaver parts had deteriorated significantly with repeated use in demonstrations, so a set of durable wax models was commissioned to supplement them.

Anna Morandi's involvement in this work came about through the artist Giovanni Manzolini, whom she married when aged 26 and still living with her parents. She had by then learned Latin and to write in a fluid style and with scientific precision. It seems she had also received some training as an artist. The project director charged with fulfilling the papal commission for a Museum of Anatomy in Bologna, in which the wax models were to be housed, was the artist Ercole Lelli. Manzolini gained a position as chief assistant to the commission, but soon clashed with Lelli over the quality of their respective models. The rift saw Manzolini resign and set up his own rival wax-modelling studio and anatomical school for medical students and interested amateurs. The facilities and work were shared with his home and wife, now aged 32 and the mother of two small boys.

The Mazolinis' studio soon established a fine reputation for quality. After her husband's death Morandi's notoriety as sole proprietor, giving demonstrations via her wax studies and the dissection of cadavers, grew to the point that Grand Tourists were flocking to see her in huge numbers. Even the Emperor of Austria visited her, and she received commissions from Catherine the Great, the kings of Poland, Sardinia and Naples, and the Royal Society. The Bolognese cultural administrators realised that it was essential to keep such a popular attraction, with all the tourist money that came with it, in the city. So to that end, they offered Morandi a small stipend and a university position as anatomical demonstrator.

Morandi kept a meticulous anatomical notebook, running to 250 pages. She begins with the eye: for twenty pages she lists and explains her models of intact and sectioned eyes. She illustrated clearly the functioning of the eye and its muscles through wax depictions of eyes in different directions of gaze surrounded by an eye with the extraocular muscles splayed out. She also modelled dissected eyelids, corneas, retinas, tear ducts, glands and nerves.

Some of her microscopic visualisations were of structures she herself discovered. The notebook description of one particular model explains that it serves to demonstrate 'how the crystalline lens is naturally so delicate that if you touch it and lightly press it with the tip of a finger, it immediately collapses and distorts.'

The mechanical aspects of eye movements in facilitating sight were of especial concern to Morandi. Of the lacrimal gland she says, 'At the ciliary puncta, located at the top of the grooves at the internal part of the tarsi, there are openings or excretors of the sebaceous gland of the bilateral form that serve the lymph that facilitates the movement of the lid and maintains the moistness of the eyeball.' Her writings display her observational and experimental skills, and show that she is not afraid to disagree with other anatomists. For instance, she writes: 'Contrary to the opinion of some authors, it is important to note that the transparent cornea is never joined to the opaque, as I have shown in all the tests I have been able to conduct.'

One of her most important original contributions to anatomy is her discovery of the course and origin of the inferior oblique muscle. In her notebook she points out the deficient current description of the muscle before adding her own, improved, version: 'The oblique inferior muscle not only attaches to the nasal apophysis [outgrowth] of the maxillary bone, as the authorities agree, but with the bone opened, one sees that the muscle proceeds and attaches to the lacrimal sac. This was discovered by me in my observations and I have found it always to be constant.'

Unfortunately Morandi's finances deteriorated rapidly. After her husband's death she was surviving largely on a meagre honorarium from the University which the officers refused to increase. Less than eighteen months after being widowed she was forced by her financial situation to give up her elder son, then aged eleven, to an orphanage. Facing financial ruin and in ill health, she accepted an offer from a local senator, Count Ranuzzi, to purchase her collection and provide her with an apartment in his palace.

Ranuzzi was a chancer purely concerned with making money and society connections. He purchased Morandi's entire collection of anatomical figures for 12000 lire in 1769. Then, in 1771, he acquired her library of anatomical atlases and texts, her dissecting and sculpting instruments and sundry other pieces for just 600 lire, to be paid in 100 lire instalments over six years. She collected only three of those instalments, as she died in 1774. In fact her bequest to her sons consisted almost entirely of the remaining payments. Ranuzzi, meanwhile, sold Morandi's library collection to the Bolognese Senate for 16000 just eight months after her death; a tidy profit.

In June 1777 the Insitute of Sciences inaugurated the Manzolini Room with a marble plaque at the entrance that stated in Latin: 'The Celebrated Works of the Anatomy of the Human Body by Anna Morandi Manzolini.' Luigi Galvani, in his speech at the opening ceremony, praised the precision and clarity of her three-dimensional models which, in his opinion, made for clearer critical tools for students than cadavers. 'These parts, true and natural,' he said, 'do not have, as cadavers do, any of the forbidding or putrid, which can nauseate or cause suffering in the tenderest of souls. On the contrary in their beauty and elegance they even stir fascination and award an almost incredible pleasure to those who study them.'

Even Galvani, though, would not admit her into the community of academic anatomists, mere woman that she was. Rather, he saw her as an outstanding artisan. There is no doubt that she was both. Her notebooks, and the 61 models now housed in the Museum of Human Anatomy of the University of Bologna attest to that. Perhaps the best tribute to her work is her own wax self-portrait depicting her, dressed as a noblewoman dissecting a brain, as the 'Lady Anatomist'.

For more detail see The Lady Anatomist: The Life and Work of Anna Morandi Manzolini *by Rebecca Messbarger (The University of Chicago Press, 2010)*

TINTS AND SPIRITS

The inventor of modern sunglasses who dabbled in spiritualism

Sir William Crookes was one of the scientific giants of the Victorian and Edwardian ages. His chemical investigations ranged across many fields, one of the last being that which immortalised his name in optics. His personal life, however, was somewhat complicated; not least because of his researches into spiritualism and his questionable relationship with a young medium.

Crookes (1832 – 1919) was in the fortunate position of inheriting a great deal of money from his father, allowing him to set up his own private laboratory in Notting Hill, London. He had studied at the Royal College of Chemistry and, after short spells as superintendent of the meteorological department of Oxford's Radcliffe Observatory and working at the College of Science in Chester, he continued to pursue his various investigations at his London base. He was a rarity for such a prominent scientist in that, other than these two brief sojourns, he never held an academic position; this, and his argumentative nature, may account for the fact that he only once published a scientific paper jointly with another named person.

His first major contribution to science was his discovery of a new element, thallium, in 1861, the first Englishman to do so since Sir Humphrey Davy's discovery of boron in 1808. Precedence for the discovery was a hard-fought affair against the claims of a Frenchman, Claude-August Lamy, with international recognition only coming finally when a Swedish mineralogist identified a rare thallium-containing mineral and named it 'crooksite'.

Perhaps even more significant were his experiments on cathode rays. Developing an earlier piece of apparatus, Crookes invented a glass tube in which a near vacuum could be created and across which a high electrical voltage could be applied, from a cathode at one end to an anode near the other end. As the gas pressure was reduced inside the tube, the stream of cathode rays would hit the end of the tube causing it to fluoresce yellowish-green. Crookes maintained that the rays travelled in straight lines; to prove this he introduced the familiar Maltese cross-shaped target into the path of the rays, which resulted in the formation of an identically-shaped shadow on the end of the tube. In 1897 J J Thompson identified cathode rays as electrons, the first sub-atomic particles to be found. The Crookes tube was also crucial to Wilhelm Röntgen's discovery of X-rays in 1895, a fact acknowledged in a contemporary *Scientific American* article.

By now Crookes was becoming part of the scientific and social Establishment. It helped that he had as a patron George Stokes, another famous name in optical theory, who, among other things, facilitated Crookes' election to membership of the influential Athenaeum Club. He was also admitted to the Freedom and Livery of the Worshipful Company of Spectacle Makers in 1899, an organisation at that time more concerned with attracting the great and the good of the City than with its optical heritage. At different times he was elected president of both the British Association for the Advancement of Science (BAAS) and the Royal Society.

As a public figure he became involved in several pressing issues of the day, such as water quality, sewage, hygiene and public health; agriculture and the fixation of nitrogen from the atmosphere (his own experiment predicted production on an industrial scale, later realised in the Haber-Bosch process); the commercial production of Welsh gold; and electric lighting. The latter subject was a matter of feverish research: could electric lighting become an economic replacement for gas lighting? Edison in America and Swan in England patented electric incandescent light bulbs. But so did Crookes. Crookes also installed electric lighting in his own house, an experiment he recounted in a letter to *The Times* (June 5, 1882) and entered the debate on electrifying the Athenaeum's lighting.

Crookes' research on ophthalmic tints was to be one of his last contributions to public and occupational health. But even in his eminence, he was still occasionally dogged by a scandal from many years previously. Initially a sceptic of the new craze of spiritualism that was sweeping the country, Crookes attended a séance after his younger brother, Philip, had died at sea. Philip appeared to speak through the medium with convincing detail (although much information had been made public during a libel trial between Crookes and the ship's captain): Crookes was hooked. He read voraciously on spiritualism and was introduced to another medium, 17-year-old Florence Cook.

Crookes was not alone in investigating these new psychic phenomena from a scientific viewpoint. But his papers in scientific journals that proposed a psychic force left him open to ridicule. Thereafter he wrote up his researches for the spiritualist audience. His initial investigations had been into a medium called Daniel Home, but his dalliance with the pretty Florence Cook led some to ask just how close his association with her was. Despite what he thought were foolproof electrical experiments (the medium was wired up to a galvanometer to detect any supposed unwarranted movement from their concealed seat), Crookes was duped by two other mediums, Rosina Showers, a friend of Cook, and an American, Annie Fay.

Scandal broke when the first two were eventually unmasked as frauds , while Fay was a known professional illusionist. Although Crookes disagreed, William Barrett, reading a paper at a subsequent

BAAS meeting, was surely correct in saying, 'a trained physical inquirer is no match for a professional conjurer.'

In 1908, due to government concern about eye injuries from glare, the Royal Society set up a Glass Workers Cataract Committee. Crookes, now in his eighties, experimented with how various metal oxides added to glass might reflect infra-red from white-hot furnaces, and photographed spectra of molten glass. He told the committee that 'it should be possible to make a glass that would be opaque to the infra-red and ultra-violet ... but hitherto I have been unsuccessful.' Only in 1913 did he find a solution that was opaque to ultra-violet and accounted for 90 per cent of infra-red, while being only lightly tinted. He realised, though, that protective tinted lenses would be useful for leisure purposes too, so tested over 300 different formulations, each numbered and labelled. For example, 'Crookes Glass 246' was the tint recommended for glassworkers. A light sage-green colour, containing ferrous oxalate with some red tartar and wood charcoal, it eliminated 98 per cent of incident heat.

The best-known Crookes tints are the A, A1, B and B2 series, all of which absorb all ultra-violet radiation below 350nm while reducing luminosity of the visual spectrum. Crookes' samples were produced at the Whitefriars glassworks in London, specialists in stained glass, and Chance Brothers, Birmingham.

An accident occurred at the latter with a batch of Crookes A glass getting contaminated with blue cullet (scraps of waste glass) resulting in a pleasant blue tint. Patented in 1926 as Crookes A2, it became very popular in spectacles. Crookes B and B2 were medium and dark shades of greenish smoke. The Crookes Alpha tint (very pale blue) is well-known to opticians; it replaced the very pale green Crookes A which was withdrawn for safety reasons due to its uranium content.

William Crookes was a remarkable example of the Victorian gentleman scientist, able to turn his hand to almost any area of research that piqued his interest. His favourite saying was, 'A man should always have a little more to do than he could possibly accomplish.' Crookes always had more to do, but he certainly accomplished much. A sunny day serves well as his epitaph, since sunglasses were his idea.

MRS PANKHURST'S SON

Did the famous suffragette neglect her myopic son's eyesight?

Emmeline Pankhurst is best known as a leading light of the suffragette movement, fighting tirelessly for most of her life for the right of women to vote in elections. But she was also a single mother of limited means struggling to bring up four children (a fifth child died, aged four, from diphtheria). And there has long been a suggestion that she cruelly denied her son the spectacles he needed to correct his poor vision.

How to reconcile the idealistic proponent of women's rights and devoted mother (who was so grief-stricken at her young son's untimely death that she could not bear to look at his likeness afterwards, locking away two portraits of him in a cupboard) with the woman who favoured her eldest daughter to the detriment of her other children and apparently lacked any sympathy for her surviving son, Harry, who was blighted by poor eyesight and other health problems?

Several sources portray Pankhurst as being, at best, indifferent to young Harry's predicament and, at worst, guilty of cruelty and neglect. Eaton, in *Well-dressed Role Models: The Portrayal of Women in Biographies for Children* (2006), while accepting that Emmeline loved her children, points out that Christabel (the eldest) was her favourite. She quotes Noble in

Emmeline and her Daughters (1971) regarding Pankhurst's failure to empathise with Sylvia and Harry as resulting from having 'no understanding of their shy and reserved natures.' Eaton says that, 'She denied Harry's need for glasses and persisted in believing, until Harry's early death, that a strenuous outdoor life would strengthen him. Noble implied that Pankhurst was too bound up in her own emotions and her admiration for Christabel to be sensitive to her other children.'

Martin Pugh, history professor at Newcastle and then Liverpool John Moores universities, claims in *The Pankhursts: The History of One Radical Family* (2001), that Emmeline ignored evidence of both Harry and Sylvia suffering from poor eyesight, refusing 'for years to allow them to wear spectacles, resulting in Sylvia enduring migraines for years' and that this was 'but one early sign of her reluctance to accept weakness in any shape or form in the Pankhurst family'. Poor Harry had endured chickenpox, measles and a fall that broke some of his teeth. Now, according to Pugh, being 'obliged to struggle on without glasses he began to experience pain whenever he read'. At school 'he found it impossible to read in the dim light'. He began to play truant, so was moved from his small private school to the local Board school that his other sister, Adela, attended.

More recently, the *Daily Mail* reviewed a biography of Sylvia, *Sylvia Pankhurst: The Rebellious Suffragette* (2012), by Shirley Harrison. The reviewer repeats the commonly accepted idea that 'Emmeline seems to have been a ghastly mother, at least to her younger children. She refused to allow poor Harry to wear spectacles, thus making it impossible for him to study at school and, effectively, blighting his life.'

There is no doubt that Emmeline Pankhurst was a difficult person to get on with. She was determined and idealistic, but so single-minded in her campaign for women's suffrage that she managed to alienate close friends, family members and political allies. She grew up in a family with a tradition of radical politics, which helped to shape her character. In 1879 she married a barrister, Richard Pankhurst, himself an advocate of women's rights. As well as supporting women's suffrage, he was responsible for the Married Women's Property Acts of 1870 and 1882 which provided for women to be able to keep assets acquired before and after marriage. His habit of championing and acting for the poor had the twin consequences of generating little fee income while alienating potential wealthier clientele so that, when he died suddenly in 1898, Emmeline found herself a single mother of four children, aged from seventeen to eight, reliant on the £200 - £300 per year she made as leader and principal speaker of the Women's Social and Political Union (WSPU) that she had helped found in 1903. Travelling the length of the country to deliver speeches espousing equality for women, staying in rented rooms or friends' houses, it is no wonder that she became a tough, forthright character, and it would be easy to believe, as several biographers imply, that she had no time for her sickly young son.

But there is another point of view. June Purvis, Emeritus Professor of Women's and Gender History at Portsmouth University, points out that the prevailing view of Emmeline derives largely from her daughter Sylvia's book of 1931, *The Suffrage Movement, An Intimate Account of Persons and Ideals*. The book is highly critical of Emmeline and Sylvia's eldest sister, Christabel, who helped run the WSPU; a sentiment no doubt coloured by resentment at her mother's favouritism towards Christabel and their differing philosophies on the way forward for the suffrage movement.

Sylvia's apparent bias against her mother is illustrated by Harry's experience of farm work. Emmeline sought advice and decided that this could provide a vocation for Harry (after previous abortive careers as a builder and secretary), in an environment of fresh country air conducive to improving his health. Alas a bladder infection forced him back to London, where Emmeline placed him under the care of a Dr Mills. In her book, Sylvia claims that her mother ignored the doctor's advice not to send Harry back to agricultural work as his constitution could not cope with its rigours but Purvis, in *Emmeline Pankhurst: A Biography* (2002), questions this. She says that 'Emmeline, in her letter of 1 April 1909, sought the advice of Dr Mills and was unlikely to have acted against it. Sylvia presents Emmeline as a neglectful mother, implying that her lack of concern for her son eventually led to his death.'

But the doctor's bills had to be paid, so Emmeline agreed to a lecture tour of America. Tragically Harry was struck down by a spinal inflammation shortly before she was due to embark. Historians have tended to follow Sylvia's view in her book that Emmeline was callous in leaving her paralysed son to proceed with her tour, ignoring the fact that she later softened her stance, admitting that Emmeline thought that Harry 'would recover as before.' Sadly the situation had worsened by her return, and Harry died in 1910, aged 21.

So was Emmeline really so cruel towards the younger Harry in denying him spectacles at school? A different perspective emerges from Adela, the youngest sister's unpublished memoir, *My Mother:An Explanation and Vindication* (1933). Referring to this, Purvis takes up the story: 'Harry, a sensitive child, developed acute and permanent astigmatism after catching chicken-pox and then measles. His Uncle Herbert, suspecting that the unhappy boy was playing truant from school, followed him one day and discovered that his nephew was slinking off to railway stations to watch the trains. Deeply worried about what to do for her son, Emmeline consulted a doctor who advised that nothing was wrong apart from the boy's nervousness and lack of self-control. She decided that it was best to remove him to another school'.

Perhaps, then, Emmeline Pankhurst was not such a bad mother after all. She did favour Christobel, and she did fall out with Sylvia and Adela, but maybe her real fault was heeding the doctor's advice about Harry's eyesight and not seeking out an optometrist.

THE WOW-WOW MAN

An eccentric physician and a recipe for meat sauce

Eccentric, eclectic, kindly: three adjectives that give a flavour of the strange life of Dr William Kitchiner. In a short life spanning 52 years, Kitchiner wrote books on cookery, economics, music, health – and optics. His works coaxed in the curious reader with their conversational tone, as if addressing a friend. And, somewhat unusually for authors of self-help guides, he actually lived by the principles that he set out for a good and happy life.

Kitchiner was born in 1775, the son of a London coal-merchant from whom he inherited a fortune of around £65,000. This gave him ample opportunity to develop his theories on living the ideal life. In character he was practical, benevolent and good-natured; qualities which suffuse his books, packed as they are with humour and common-sense advice.

He is best known for *The Cook's Oracle*, which, published in 1845, became a bestseller on both sides of the Atlantic. It was a handbook of economical cookery for ordinary families and contained, among other things, many recipes for sauces and ketchups of his own devising. One of these was the spicy concoction he called 'Wow-Wow sauce'. Incidentally, fans of the *Discworld* novels may know of Terry Pratchett's explosive parody with the same name. Kitchiner's public was presumably prepared for the gastronomic delights of *The Cook's Oracle* by his 1822 work,

The Art of Invigorating and Prolonging Life, although this book sensibly ended with a chapter entitled *The Pleasures of Making a Will.*

Of particular interest to the optical fraternity is his *Economy of the Eyes* (1824). The title continues: *Precepts for the Improvement and Preservation of the Sight. Plain Rules Which Will Enable All to Judge Exactly When, and What Spectacles are Best Calculated for their Eyes.* It is a compendium of useful information for the layman but provides incontrovertible proof that the average patient's vision-related expectations and misunderstandings have not changed in nearly 200 years. Who, for instance, has not had a similar experience to the following:

'When persons apply to an Optician for Spectacles to read or work with; they should clearly understand, that the Objects for which such Spectacles are solely calculated, are not placed more than 12 or 14 Inches from their Eyes.......for there seems to be a natural impulse in most persons, that after a printed Book has been handed to them for trial to read, they will presently look off – to some object on the other side of the Room, or across the Street, and say, 'Why now I can see well enough to Read with these Glasses – but I cannot discern what that word is over yonder Door;' and the Optician has oftentimes no little trouble to convince them, that such Spectacles are not intended to show Objects at a distance – to see which, their Sight is as strong as ever; and in fact, that they can see distant objects best with their naked Eye.'

And how true it still is sometimes that, 'After persons have used the same glass for some years – and it is broken &c. it is often extremely difficult to make them think, that any new one suits their Sight exactly so well as the Old one which they had been in the habit of long using.'

Kitchiner also recounts a case related to him by an optician friend of a somewhat muddled (although logical in her own way) old lady of the type that many practitioners must have encountered. This lady had consulted many eminent opticians about making reading spectacles; yet no prescription could be found that could improve her vision satisfactorily. Where others failed, the friend determined to succeed. Several pairs of spectacles later, however, the lady complained, 'No, not one of these will do – I can see better with my Naked Eye. Well!! what an unfortunate creature I am, at my Age, not to be able to read in Spectacles!!' He noted consolingly that many a person at her advanced age had difficulty seeing at all, let alone reading; whereupon the astonished practitioner was interrupted by a vehement outburst. 'Sir,' she exclaimed, 'you are strangely mistaken, Sir! I did not tell you I could not see to read, Sir! I can see to read, Sir, as well as ever I could. I only complained that I could not see to read in Spectacles!! *I can see to read very well without!!!* But my Acquaintance say how charmingly they can see with Glasses, and surely it is very hard that I cannot enjoy the same Advantage.'

The internet, and its effect on spectacle dispensing, may be a modern phenomenon but, as if to show that there is not much that is truly new in the world of optics, Kitchiner includes in one of the book's appendices *Dr Smith's Rules For Choosing Spectacles*. The citation, from Smith's *Optics*, begins: 'In order to determine the properest Glasses for defective eyes, the distance from the eye, where an object begins to appear confused, should be found – by measuring the least distance from which a *Long-sighted* person can read a newspaper distinctly and readily: and likewise by measuring the greatest and least distances from which a *Short-sighted* person can read small print readily.'

Then, '.... any person may be fitted with the properest Glasses though he lives at a distance from the shops where they are sold, by sending their focal distances computed by the foregoing rules.' Not so different from emailing one's prescription and receiving the spectacles by post?

The Economy of the Eyes also discourses widely on both opera-glasses and telescopes. An enthusiastic amateur astronomer, Kitchiner kept several fine specimens of the latter in his home at 43

Warren Street, Euston. The hospitality he provided for his friends here was legendary. Invitees to his famous dinners were requested to dine with the 'Committee of Taste.' He would select and prepare the food himself, and punctuality at dinner, in order to protect 'the perfection of the several preparations....so exquisitely evanescent that the delay of one minute, after the arrival of the meridian of concoction, will render them no longer worthy of men of taste' was insisted upon. Dinner was at five, supper at seven-thirty and at eleven he retired to bed.

On 26 February 1827 Kitchiner attended a dinner given by the noted singer, Mr Braham. Unfortunately he enjoyed himself a little too much, ignoring his own rule staying later than usual. Once home he was taken ill and was dead within the hour. Here, as a final tribute, is reproduced his recipe for Wow-Wow sauce for stewed beef:

Chop some parsley leaves very fine; quarter two or three pickled cucumbers, or walnuts, and divide them into small squares, and set them by ready; put into a saucepan a bit of butter as big as an egg; when it is melted, stir to it a tablespoonful of fine flour, and about half a pint of the broth in which the beef was boiled; add a table-spoonful of vinegar, the like quantity of mushroom ketchup, or Port wine, or both, and a tea-spoonful of made mustard; let it simmer together till it is thick as you wish it; put in the parsley and pickles to get warm, and pour it over the beef; or rather send it up in a sauce-tureen.

4. Sport and Miscellany

PRECISION OPTICS AND A WELSH FAIRYTALE

The rise and fall of an optical factory's works football team

In 1981 a plucky little football team from south Wales, recently promoted to the Third Division of the Football League (equivalent to today's League One), found themselves in the quarter-finals of the European Cup Winners' Cup by virtue of having also won the Welsh Cup the previous season. Barring their way to the semi-finals were the three-time East German champions and runners-up on nine occasions more. Could Newport County pull off the shock of the season? And what does a football match have to do with precision optics?

Newport's opponents were FC Carl Zeiss Jena. They were formed in May 1903 by workers at the Carl Zeiss optical factory and, sponsored by the company, originally went by the name of Fussball-Club der Firma Carl Zeiss, although the club's name was to undergo several changes down the years. With the growth of football as a popular pastime in the late nineteenth century it became common for employees of large manufacturers to run sports teams. A team formed by workers in the Carriage and Wagon department of the Lancashire and Yorkshire Railway became Newton Heath LYR FC, which later underwent another name change to Manchester United. In the Netherlands, Philips actually formed a club for the benefit of its employees, Philips Sport Vereniging, better known as PSV Eindhoven.

The Zeiss works originated with Carl Zeiss (1816 – 1888) who opened his first workshop and small store in 1846, in Jena, making and selling optical instruments. At that time building optical systems for microscopes was largely a time-consuming trial-and-error affair. In the 1860s Zeiss engaged the help of a physics lecturer at the University of Jena, Ernst Abbe, to help put lens construction on a more scientific footing. Abbe's work in the subject and the formulas he derived enabled the Zeiss microscopes to gain an international reputation for quality; and in 1876 (with retroactive effect from 1875) he became a partner in the firm.

Abbe was also quite a social reformer. He had become sole director of the company on Zeiss' death, having acquired his heirs' shares. Perhaps his most important act was to establish the Carl Zeiss Foundation in 1889 in order to ensure the continued existence and ethos of the company irrespective of personal ownership interests. This he did by transferring all the Zeiss shares to the foundation, making it sole owner. He also gave it his shares in the Schott specialist glass-making company that worked closely with and supplied Zeiss; the founder, Otto Schott, later transferred his shares in 1919, making the foundation sole owners of the glassworks too.

The employees' welfare was also of great concern to Abbe. The foundation enshrined legal obligations of the management to maintain good labour relations. He introduced profit-sharing, a right to retirement pensions, sick pay and holiday pay. In 1900, perhaps remembering the hard life of his father as a fourteen-hours-a-day foreman in a spinnery, he brought in the revolutionary concept of an eight-hour working day. It is fitting then that the football club, having commenced building its stadium in 1921, decided to rename it for its fifteenth anniversary the 'Ernst Abbe

Sports Field', an arena presently with a capacity for just under 13,000 spectators.

There was a fair amount of success for FC Carl Zeiss Jena (CZJ) in its early years, winning the regional league twelve times during the period 1910 – 1933. The Nazi regime and partition of the country after the Second World War inevitably created upheavals in the organisation of German football, but CZJ also flourished in the post-war years in the East German League, reconstituting as a club independent of the Zeiss company in 1966. Several of their players were capped by East Germany, notably Konrad Weise, their most capped player (86 caps), who played for the club between 1970 – 1986. Perhaps their most famous son is Bernd Schneider, who won 81 caps for the unified Germany, although his success came at Bayer Levekusen (another company employees' club, incidentally).

At the end of the 1979-80 season CZJ finished third in their league and were about to become runners-up for the ninth and last time: one of the powerhouses of East German football. Newport County had just been promoted to the third tier of the four-division English league for the first time. Both had qualified for the following season's European Cup Winners' Cup (CWC) by winning their respective country's domestic cup competitions (with the CWC now defunct, the clubs would nowadays be entered into the Europa League). CZJ might be expected to progress well in the competition – but Newport? A relatively kind draw in the first two rounds saw Newport beat Crusaders of Northern Ireland and SK Haugar of Norway 4-0 and 6-0 respectively on aggregate, while CZJ successfully navigated a much more difficult route past Roma and Valencia.

CZJ and Newport were drawn together in the quarter-finals. The first leg of the tie was at the Ernst Abbe Sports Field on March 3, 1981. As if Newport's task wasn't difficult enough, they were without their striker John Aldridge, later to star with Liverpool, through injury. Predictably the Welsh minnows went behind early on, but managed to equalise. A late goal by Jürgen Raab, scorer of both goals, put CZJ ahead again; but an even later goal, in the 90th minute, by Tommy Tynan, also his second of the match, enabled Newport and their 200 doughty supporters to return home from behind the Iron Curtain with a 2-2 scoreline.

That season Newport had an average home attendance of about 5,600. On the night of March 17, their Rodney Parade ground was heaving with 18,000 people eager to witness the Welsh defend their two away goals. But inspired goalkeeping and goal-line clearances frustrated Newport, while a converted free-kick gave CZJ a 1-0 win on the night, sending them through to the semi-finals 3-2 on aggregate. CZJ squeezed past Benfica 2-1 on aggregate in the semis, but lost by the same score in the final to Dinamo Tbilisi.

That period proved somewhat of a watershed for both clubs. Newport enjoyed a few seasons of domestic success but in 1987 they were relegated from the Third Division, and dropped out of the League the following season. German reunification proved tough for CZJ, as it did for most East German clubs, finding themselves pitted against wealthier West German outfits. Initially placed in the second division of the new Bundesliga, CZJ now play in the fourth tier of the league pyramid. Newport have at last finally turned their fortunes around. Under the management of ex-Tottenham player, Justin Edinburgh, Newport in 2013 won promotion back to the Football League after 25 years' exile.

The same year, when CZJ celebrated the 110th anniversary of their formation, they invited their old adversaries over to play a friendly match in remembrance of their famous CWC tie. Newport-based group, *Flyscreen*, who wrote a song at the time, called *Carl Zeiss Jena*, to celebrate the 1981 matches, re-recorded it and released the new version to mark the anniversary and re-match. On July 13, a crowd of 2800, including 250 from Wales, saw CZJ take a two-goal lead. Newport pulled one back and, as if in commemoration of that famous first leg match, gained an equaliser right at the death to replicate the finale of the original 2-2 draw. Who knows when and in what circumstances they will meet again?

EYES ON THE ASHES

Optical connections of the greatest cricketing rivalry

'The body will be cremated and the ashes returned to Australia.' So ran a mock obituary for English cricket in the *Sporting Times* after England lost to Australia on home soil for the first time, on 29 August 1882, at the Oval. One of the oldest international sporting rivalries, begun in 1877, now had a name: *The Ashes*. As England, the current holders, do battle this winter to retain the urn, we look at some of the figures in Ashes history who have particular optical connections.

Bob Woolmer, former Ashes player for England and respected international coach, wrote a monumental distillation of current scientific and practical cricket knowledge which was published just after his untimely death during the 2007 World Cup. In *The Art and Science of Cricket* (New Holland, 2008) he discusses at length the visual problems that batsmen must overcome when facing different types of bowling.

Regarding what makes the best batsmen, he concludes, 'But at the core of their success is the ability of their subconscious brains to process visual information, available to all, more accurately and more rapidly than others.' And he cites the pithy summary of former Australian captain, Greg Chappell, in his *The Making of Champions*: 'The brain is a better cricketer than you'll ever be.' But what if the same visual information is not available to all?

Bill Ponsford (Australia 1924 – 34)

Bill Ponsford was a contemporary of Don Bradman, the greatest of all batsmen. Bradman averaged 99.94 in Test matches, the next best in history being 60.97. Ponsford's Test average was a still very decent 48.22, but then he was severely colour-deficient. A discussion of his probable protanopia can be found in *One of Cricket's Greats: A Protan Mystery* (D Baker, *Optician, 09.11.07, pp32-4*). He certainly did not have the visual information 'available to all' but how much of a difference did it make to his ability to sight the ball? Once, when asked, he replied, 'I never noticed its colour, only its size.' How much better might he have been with normal colour vision? The question is moot, but the English players of the time would no doubt be thankful that they did not have to cope with another Bradmanesque batsman in the same Australian team.

Geoff Boycott (England 1964 – 82)

Geoff Boycott is not only one of the most famous cricketers in history; he was also one of the most iconic spectacle-wearing players – until he switched to contact lenses in 1969 (a Kelvin contact lenses showcard showing Boycott batting in lenses was produced in the early 1970s). That season started well for him with his county, Yorkshire, but a run of three ducks in four Test innings put a dampener on it. During the 1977 Ashes Test at Trent Bridge, he was at fault for running out his partner, Derek Randall, yet composed himself enough to complete a century. In the next Test, at his home ground, Headingley, he scored his hundredth career hundred and became one of the few players to be on the pitch for every minute of a match. Typical Boycott. On his Test debut (at that time wearing spectacles), also against Australia, there is a story that the Australian skipper, Bobby Simpson, called over to the bowler, Graham MacKenzie, 'Look at this four-eyed ****. He can't **** bat, knock those ****glasses off him straight away!' Only Boycott knows if that incident had any bearing on his subsequent switch to contact lenses.

Bill Bowes (England 1932 – 46)

Another bespectacled, quite myopic, cricketer. Standing at six-foot-four, ungainly-looking, he couldn't bat or field, but he was a fine fast-medium bowler for Yorkshire and England. He played in two home Ashes series, but managed only one Test in Australia, on the 'Bodyline' tour of 1932-33. Bodyline, or leg-theory, was a strategy of bowling short at the batsman's body to a packed leg-side field, developed by England captain Douglas Jardine expressly to combat the genius of Don Bradman. Bowes' most famous wicket came in the second Test, at Melbourne, in December 1932. Bradman had missed the first Test, and was greeted with thundering applause on coming out to bat. So much so that Bowes had to interrupt his run-up to wait for the cacophony to subside. During this pause he decided to move one of his fielders. A second attempt at running in also had to be aborted as the applause continued, so he made another field change while waiting. He noticed that Bradman had observed these changes carefully, which would have suggested to him that he was about to receive a bouncer. So Bowes bluffed Bradman: instead of pitching the ball short, his delivery was slightly slower and fuller. Bradman, already in position to hook the anticipated bouncer, had to adjust his shot and, in doing so, played the ball onto his stumps. Bradman, out first ball for nought!

David Steele (England 1975 – 76)

David Steele only played in one Ashes series. But he made a huge impact at the time, and the image of this grey-haired, bespectacled, batsman has become part of Ashes iconography. Looking much older than his 33 years, one journalist dubbed him 'the bank clerk who went to war.' England had been thrashed in the First Test by an innings and 85 runs, precipitating the resignation of the England captain, Mike Denness. The new incumbent, Tony Greig, plucked Steele, a journeyman batsman from Northamptonshire with a modest record, from relative obscurity. Greig sensed in Steele the toughness to face the twin fast bowling menaces of Lillee and Thomson, then in their pomp. The second Test was at Lord's. On going out to bat, Steele famously descended one flight of stairs too many in the pavilion, almost ending up in the lavatory. But he made a gritty 50, and followed this up with three more half-centuries in the series. He could not prevent a series defeat, but a point had been made about facing up to intimidating bowling. Such was the effect of his efforts against Australia, he was named BBC Sports Personality of the Year 1975. And as a cricket coach at Oakham School, he has had a hand in nurturing Stuart Broad, one of England's current stars.

Geoff Lawson (Australia 1980 – 89)

Geoff Lawson has many strings to his post-playing career including broadcasting (he comments on cricket on television, radio and in print), public speaking, fundraising, coaching and university and sports administration. He is also a qualified optometrist, graduating from the University of New South Wales in 1984. Lawson announced himself to the cricketing world (and endeared himself to fellow Australians) by hitting and then dismissing Boycott as a bowler for NSW playing against the England XI in 1978. Fate put him on the receiving end ten years later when he had his jaw broken by Curtly Ambrose, playing against West Indies in a Test at Perth. In all, he took 180 Test wickets, 97 of which were against England. As captain of NSW he espoused a playing philosophy of 'get on or get out' which influenced heavily NSW batsmen and future Australian captains, Mark Taylor and Steve Waugh. They led phenomenally successful teams and were all-conquering in Ashes series. In fact, through their adoption of Lawson's aggressive batting policy, it can be fairly said that Test cricket was completely revolutionised as a faster-scoring, higher-tempo game. One wonders: will there ever be another optometrist Test cricketer?

THE SQUASH BALL IN COURT (PART I)

What colour should a squash ball be?

The humble squash ball is just 40 millimetres in diameter and weighs 24 grams. It hasn't changed much since a standard version was agreed upon in 1923. Discussions of technological developments in the sport centre mainly on the racket. Yet there are two innovations to the squash ball concerning its visual attributes that have posed interesting issues of patent law. The first, concerning a new ball colour, is the subject of this article. The second, a reflective ball, will be covered in the next piece (see page 194).

Squash was born of a variety of racket games, but popularised by the boys of Harrow School in the mid-nineteenth century. Traditional rackets balls were too bouncy for the confined spaces in which the boys played, but it was found that a punctured rackets ball or balls of softer material 'squashed' more on impact with the walls allowing for more variety and skill of play. Gradually playing conditions became standardised, although by the early 1920s there were several types of ball in use.

The Royal Automobile Club built three squash courts in 1912 at its premises in Pall Mall and subsequently became prominent in the development of the sport. One RAC member, Col R E Crompton, carried out extensive research and testing on a variety of balls then available, as a result of which the club adopted their preferred ball. A meeting of delegates of squash-playing rackets clubs in 1923 agreed that the RAC ball should become the standard for the sport.

Traditionally squash balls have been made from rubber mixed with synthetic additives to give them the required qualities of strength, resilience, colour and curability (the ability to be cured, or vulcanised). Two half shells are moulded which are then glued together and buffed to give the characteristic matt black finish. Not much changed until the 1970s, apart from some experimentation with a synthetic alternative to rubber, that was unsuccessful due to the tendency of these balls to split.

There was one longstanding disadvantage with the black rubber ball, however: it marked the white squash court walls. A non-marking dark green ball was developed with some success but, in 1975, a patent application was filed for a new blue squash ball which, it was claimed, gave a surprisingly beneficial visual advantage to players. At issue was whether this 'discovery' amounted to an 'invention' ('manner of new manufacture') as required by the Patents Act 1949 (which was then in force). This involved considering whether the new colour provided an unexpected or non-obvious technical benefit over the 'prior art'; and whether this identified a problem with the prior art even though the solution, once found, seemed obvious. Essentially, was the new colour of an incidental, ie cosmetic, rather than technical nature?

The judge hearing the case in the Patent Office pointed out that in many sports balls are coloured to enhance visual contrast, so the applicants couldn't generally claim an invention by colouring a limited class of ball; neither the mode of manufacture nor the advantage sought was new. But the applicants were seeking to establish that their ball had an unexpected, unusual or non-obvious merit. This, said the judge, was tantamount toclaiming that a blue/white combination was not one that would naturally be considered for contrast purposes; on the contrary, many traffic signs, for instance, used just that combination.

Evidence for the visual advantage produced by the blue ball was offered in the form of testimony from a noted squash player who had been asked to try it out. But the judge noted that such testimony was particular and could not necessarily hold for all players, neither could it be extrapolated for all playing conditions. The judge's opinion was that the applicants had not established a uniform advantage over existing squash balls; they offered only an alternative. As such, there was an inherent lack of novelty: the claims related to a known squash ball given a particular colour for 'eye appeal', so the application was denied.

The applicants appealed to the Patents Court in July 1978. The judge there was impressed by new evidence supplied by the inventor. Having experimented with many colours, he had settled on a colour denoted as 'Flander's Blue', a sort of royal blue, defined as shown in the British Colour Council (BCC) Dictionary of Colour for Internal Decoration. The Dictionary is the BCCs standard colour range for paints, comprising around 60 colours, arranged in a logical order using Munsell notation.

The Munsell Book of Colour defines colours by assigning them values of three attributes: Hue (colour), Value (lightness factor) and Chroma (intensity). As well as the above-mentioned squash player's testimony - '... much to my surprise and pleasure ... the ITS blue ball enhanced the acuity of my vision of the fast moving ball' - the inventor produced questionnaires completed by other squash players that corroborated the effect of surprisingly good visual perception of the blue ball.

Further evidence was presented of the commercial success of the new ball: sales of over a quarter of a million in 15 months; other manufacturers marketing their own versions; and advertisements and press releases in sports magazines. The judge pondered as to why, since sufficient novelty in the product was the issue, earlier experimenters with non-marking balls had not happened upon this blue colour? The above evidence would indeed suggest that there was a distinct advantage to the blue colouration, negating the argument that the blue ball was merely 'a known ball of particular colour chosen for eye appeal.' Hence the application did relate to an invention (a 'manner of new manufacture'), and the appeal was allowed.

The result of the appeal was a surprise to some patent law experts, as it seemed to be a weakening of the principle that mere discoveries and cosmetic changes to an existing idea are not patentable. As *New Scientist* of 29 March 1979 reported in its technology pages, 'Although BP 1 538 860 was granted under the old patent law, the case must surely influence future attitudes to what is permitted within the new laws.' Even though patent law has evolved since then, this case is significant in that it is still often quoted to illustrate the general principles of technical versus cosmetic developments.

The general idea can be summarised as follows. A dark green ball is developed: is it patentable? Were the only difference from the black ball cosmetic – no. But as it solves the technical problem of marking the walls, the answer is yes. A blue ball is then developed: is it patentable? The technical problem of marking has been solved so, again, if the only difference is cosmetic – no. But it solves an 'unexpected' technical problem of visibility, so (as was argued in the appeal) the answer is yes.

The visual attributes of the blue ball have in part been obviated by newer technologies (as discussed in the next article). The only blue ball currently in production is the Dunlop Max, a ball 14% larger than standard, designed for the complete beginner. But optical professionals are entitled to ask: why was there no consultation of vision specialists in the patent case? As the original judge remarked, '..."chance" findings [of enhanced visibility] ... may not be so surprising at all if the problem to be solved is put on a proper research basis ...'.

THE SQUASH BALL IN COURT (PART II)

A reflective squash ball and a legal quandary

The technical problems of squash ball visibility may not comprise the largest area of vision-related sports engineering science, but they have produced two cases of interest to patent lawyers that are still cited regularly as evidence in forming legal decisions. The last piece (page 190) dealt with the invention of a blue squash ball; this one looks at an innovative solution to the problem of improving visibility of the ball for a television audience.

The main legal interest stems from the problems raised by an invention, the result of collaboration by two colleagues; their proportional input into the two essential elements of which it consists, and thus their relative entitlement to the property in, and grant of, a patent for it. Of interest too is the pre-existing patent relating to one of the elements, and whether this affected the entitlement of the colleague whose contribution could be considered effectively 'prior art'.

Before diving into the story of the invention and unpicking the legal arguments surrounding it, it is worth describing the invention itself. The 'reflector ball' or 'teleball' had the features of small pieces of retroreflective material applied to recesses in the surface of the ball. These recesses, which would protect the reflective material, were similar to the one that squash balls have long had, within which is the coloured dot that denotes ball speed.

The actual patent application set out several possible ways of achieving this, including the types of, and methods of applying, the reflective material; and the option of having a single reflective layer below the external surface of, possibly, a transparent ball.

The genesis of the idea was a series of discussions in the early 1980s between two dentist friends, Messrs Christie (a keen squash player) and Godin, regarding the possibility of manufacturing a reflective squash ball to enhance televised visibility. A patent application was initially made by Godin alone, which eventually lapsed. An application was made latterly by Christie, which Godin referred to the Comptroller of the Patent Office under section 8(1)(a) of the Patents Act 1977, claiming either sole or joint entitlement to the property in and grant of a patent for the invention.

The case was heard by the superintending officer for the Comptroller. He determined that the essential elements of the invention were the use of reflective material and its positioning below the external periphery of the ball. The onus was on the referrer, Godin, to establish, on the balance of probabilities, that he had made contributions to these elements.

A complicating factor was a prior US patent for the use of reflective material on sports balls, available publicly since 1977 in the Science Reference Library (now part of the British Library), of which apparently both parties were unaware. The examiner ruled that this was not relevant in the context of which of the ex-friends was responsible for the idea, as the term 'invention' in the Act could refer to an idea at the time of application even if it turned out to be non-patentable. But the British Library now refers to this case in its brochure of services to illustrate the importance of its date stamping system, as applied to every document it receives.

Turning to the actual matter of whether the parties were singly or jointly responsible for the invention, the examiner found that the idea of using reflective material was largely Godin's contribution: he had sourced various materials, investigated their use and made the first prototype reflective ball by himself. As for positioning the material below the surface, and in particular, in recesses, the examiner held that this had most likely come from Christie. Christie, the keen squash player, formed the idea for protecting the material from scuffing and degradation this way from his knowledge of the existing recess in squash balls housing the coloured speed identifier. Also, one of Godin's prototypes had reflectors stuck to the surface, suggesting to the examiner that, at the time, he attached no great importance to recessing them.

The examiner's decision was that all possible versions of the reflective ball as set out in the application – excepting that of the transparent ball with sub-surface reflective layer, which was assignable to Godin alone – were jointly owned by the two parties. An appeal on the basis that the versions specifying particular ways of recessing the reflectors, as determined by Christie, should belong to him alone was dismissed.

A principle was established that a claim embodying both parties' ideas together with an additional feature contributed by one party was nevertheless jointly owned. (For those wishing to learn more about the case, the appeal was filed in the name of Viziball Ltd, as Christie had assigned his patent application to that company.) Questions of priority and claim to inventorship that occur regularly with patent applications make this a much-cited case in patent law.

One of the other features of the patent application was the televisual system. It described a method whereby the path of the ball would be illuminated and a camera would be directed towards that path to receive images of the light reflected from the reflective material in the ball. In practice this required a beam of light directed towards the court from behind the back wall, where the main camera would be. Televised matches of that period used a court with a glass front wall, behind which Press photographers would be sited in order to get the best shots of play. They found that having a bright light beam pointing straight towards them destroyed any hope of getting a decent picture.

In any case, the reflector ball's demise was already assured courtesy of another, slightly earlier, invention that was gaining ground in the sport. The logical extension of a court with a glass front wall was an all-glass court. Vision control panels, which allowed for plain, single-coloured one-way transparent walls became available in the late 1970s and made the all-glass court viable. The panels, produced by Contra Vision Ltd, have a partially imaged substrate and a design superimposed on a print pattern. This pattern can be opaque, as with the squash court, in order to achieve one-way vision panels incorporating the court's wall markings. Such courts have walls that are opaque to the players but, with sufficient internal illumination provided by overhead lighting, are simultaneously transparent to outside observers. The conditions are ideal for television cameras and allow for seating to be arranged around all sides of the court rather than solely behind the back wall. Photographers can be sited comfortably, undisturbed by unwanted light, at their preferred location facing the front wall.

The squash ball started life black. It became green in order to reduce the markings made by the ball on the squash court walls. Battles have been fought in the Patents Court over, firstly, the introduction of a blue ball alleged to improve visibility and, secondly, a reflective ball supposed to do the same, especially for televised matches. The blue ball disappeared virtually without trace, and the reflector ball was superseded by another technology. Almost all squash is played nowadays with a traditionally-coloured black or green ball. A footnote to the story is that the visual conditions created by the opaque (from the inside) walls of modern all-glass courts have led to fairly successful experiments in televised tournaments with another new ball. Its colour? White.

A ROMAN THERAPY

Animal dung and other optical cures

One would naturally tend to associate animal dung with a cause of eye diseases rather than their cure. But this was not always the case. A tour through the pharmacopoeia of ancient Roman times reveals uses for a host of strange ingredients. Some are quite logical in the light of modern medical knowledge. Others seem bizarre, and a few, potentially poisonous. It is not for the squeamish.

The delivery method of choice for ocular medicaments in Roman antiquity was the collyrium, a latinised Greek term commonly used to mean an eye ointment. It could be liquid, where powdered ingredients were mixed with solvents, or dry, in the form of ointment sticks. These sticks could be applied directly if necessary or preserved so that a piece could be broken off for use when required, perhaps to be ground up or made into a solution. To facilitate the latter, it would be triturated with water, milk, breastmilk, egg-white, wine or vinegar, and applied as drops or ointment. There is abundant information available on the ingredients and preparation of collyria via the extant writings of contemporary physicians and the collyrium seals that have been discovered over the last 350 years or so throughout Europe at Roman encampment sites.

The most famous contemporary source is Galen (130 – 201AD). He wrote a complete *pharmacopoeia oculorum*, which contains over 200 collyrium prescriptions, often giving precise instructions for preparation, always explicit on the area of application and usually with an explanation of the method of application and treatment. Another important source is Dioscorides (40 – 90AD). His work, *De Materia Medica Libre*, provides exact information on the effect of various drugs and mineral substances on eye diseases. It was so influential that it was widely used in Europe up until around the end of the sixteenth century, and originated the term 'materia medica' for a collected body of knowledge about medicinal substances. The extent and thoroughness of his medical writings is praised by Galen: '... Dioscorides from Anazarbus has, in five books, written on the collected teachings of medicine in a useful way, in that he has not only covered herbs, but also trees, fruits, natural and artificial juices and furthermore, metal and animal substances. To me he seems to have covered the teachings of medicine to a greater extent than anyone else.'

Over fifty distinct classes of ingredients of unearthed collyrium seals have been identified (the seminal work *Ancient Ophthalmological Agents*, by Harald Nielsen, Odense University Press, 1974, was one of the first to provide a comprehensive breakdown of the composition of various collyria). It is notable that, of Nielsen's thirty-nine collyria examined, certain ingredients were present within the majority. These are *cadmia*, *croceum*, *opium*, *piper* and *gummi*.

Cadmia, the most common compound found in collyria, has been identified with zinc oxide. Dioscorides describes how the substance is produced by the deposition of smoke on oven walls during the process of melting down copper in mines. He further elaborates on the preparation of *pompholyx*, which he refers to as 'smelting-furnace smoke' produced by mine workers throwing finely-crushed *cadmia* into the smelting furnace to improve the metal. The resultant smoke product was white and greasy. Zinc is common in copper ores, so these descriptions almost certainly describe zinc oxides. The physician, Plinius, describes the efficacy of *cadmia* on epiphora, caligo (a form of corneal opacity) and aspritudo (trachoma). It is worth noting that a completely pure zinc oxide preparation was not made until the seventeenth century, by the physician, Jean Rey. The ancients sometimes added wine or vinegar to their preparations, which produced a measure of zinc acetate as well.

Croceum is saffron, derived from the dried styles of the *Crocus sativus* plant. In ophthalmology it was mainly used for epiphora. Explains Dioscorides, 'In fact it is diuretic and a little astringent, which is why, together with water, it works as an ointment against erysipelas [an acute streptococcal skin infection] and against pus from the eyes and the ears.'

Opium, known as *lacryma papaveris*, the latex of the opium poppy, was known as a painkiller from early times. The method for obtaining the drug is found as early as 380BC in the writings of the Greek physician, Diogeras. Dioscorides explains precisely how the incisions should be made in the flower capsule so as not to spill any latex. The method is exactly the same today.

Three forms of *piper* are described. Although known to be forms of pepper, their origins were a matter of debate since they were obtained from remote locations. Nevertheless they were considered valuable as drugs and often used in the treatment of eye disease. It was Matthiolus who, in the mid-sixteenth century, expounded the correct origins of *piper longum*, *piper nigrum* and *piper album*: the former being the product of Piper longum and Piper officinarum plants, and the latter the unripe, dried fruit and the ripe fruit, respectively, of the Piper nigrum plant.

Gummi was a form of gum arabic, obtained from the plant Acacia Senegal, a species of Mimosa. It is ubiquitous in collyria prescriptions because of its use as a binding agent. The ancient Egyptians were well aware of this substance, which they called *komi*, as is evidenced by inscriptions on monuments from the seventeenth century BC. Acacia itself was used as an eye medicament.

Many other of the listed collyrium ingredients are plant or mineral in origin. Often their uses would seem reasonable to modern medicine, eg opium (pain relief), aluminium compounds (astringent); and some downright dangerous, eg toxic antimony compounds. But the animal-based ingredients are perhaps the most curious. *Castoreum* was thought originally to be cut-off and dried beaver testicles. Plinius asserts that the beaver knew why it was being chased and would bite off its testicles when in a position of danger. Dioscorides says that this must be false as the testicles 'lie flat, hidden under the skin as on the pig.' In the fourteenth century it was discovered that *castoreum* was prepared from beaver glands to be found under the skin between the lower part of the rectum and the sex organs. Cut out and dried, and mixed with attic honey, the preparation was used to cure blurred vision.

Fell tauri is bull's gall, but the gall from many animals (and humans) had their medicinal uses. The gall of the fish, kalliomynos, was especially good; fish-gall is mentioned as especially effective for eye complaints in the Book of Tobit in the Apocrypha. Wild goat's gall was thought excellent for muscle paresis-induced squint. Not only was hyena gall used as eye medicament, but merely rubbing an inflamed eye with a hyena paw was said to be effective.

Crocodilonium stercus, or crocodile excrement, often appears in eye remedies. Plinius says that, mixed with leek juice, it is very effective against eye diseases. Dioscorides goes to town with excrement, listing that from oxen, dogs, sheep, goats, donkeys, pigeons, boars, hens, storks and mice as being effective in medicine, although not all for the eyes. Galenextends even this list but highlights goat excrement as useful for rough eyelids. Through Galen's great influence down the ages, these medicaments remained in materia medica into the Renaissance.

In conclusion: a pharmacological treatment for cataract? It's been done. Here, paraphrased, is the circa-fifth century Byzantine physician, Aetius's, prescription for cataract. A whole, living, male adder is placed in a pot with some fennel juice and frankincense.

The pot is smeared on the outside with mud and then burnt until the contents become ashes. Grate them and use as a dry medicament.

THE MALVERN CURE

The healing waters of England

Advertisers have been fond of the slogan, 'Nothing works better than...' to which the obvious cynical response is, 'Well, take nothing then!' The town of Malvern, nestling in the lee of the Malvern Hills on the western edge of Worcestershire, could have been designed for that response. Malvern is best known as a spa town, but its water, which filters through the limestone and granite rock of the hills to a number of springs at a collective rate of about 60 litres per minute, is unique among English spas because of its purity. And one of those springs has long been famous for its curative properties for eye disease.

A priory was founded at Malvern in 1085, whose monks reportedly bathed in the Holy Well situated in the hamlet of Malvern Wells. The first definite record of this well dates from 1558 with the granting of the spring to John Hornyold by Elizabeth I. About 200 metres higher up a slope is sited the Eye Well, so called because of its fame in curing eye diseases. A 1662 work by the physician and oculist Richard Bannister (sometimes spelt Banister) noted this property. His book, known as *Breviary of the Eye*, fully titled *Treatise of One Hundred and Thirteen Diseases of the Eyes*, was actually a second edition of a translation of a work by the French physician, Guillemau, with some additions by Bannister. One of the additions was a verse about the Eye Well:

'A little more of their curing tell,
How they help sore Eyes with a new found Well.
Great Speech of Malvern Hills was late reported,
Unto Which Spring People in Troops resorted.

The founder of Malvern spa was a physician at the nearby Worcester Royal Infirmary, Dr John Wall (1708 – 60). He became interested in the properties of Malvern spring water and conducted the first analysis of it, comparing it with spa water from Bath, Cheltenham, Bristol, Scarborough, Tunbridge Wells, Worcester and the original spa town - Spa, in Belgium. The results, published anonymously in 1754 by Wall in *Experiments and Observations on the Malvern Waters*, found that Malvern water contained 'the least amount of dissolved matter'; the geology of the area, described later, explains why. Indeed, Wall concluded that 'the Efficacy of this Water seems chiefly to arise from its great Purity.' A local rhyme immortalised that opinion thus:

The Malvern water says Doctor John Wall
Is famous for containing just nothing at all.

And effective the water certainly seemed to be. Wall's short book was so popular that he compiled an appendix of 51 cures, 47 from the Holy Well. One cure, from the Eye Well, concerned a woman from Bewdley, near Kidderminster, who had been afflicted with such a severe eye infection for nine months that the purulent discharge had sealed her eyelids together rendering her sightless. All the treatments tried had at best given her only temporary respite. On Dr Wall's advice she travelled to Malvern and, after bathing her eyes for a week with the Eye Well's waters, her sight was 'so recovered that she could see a Flea leaping on her bed.'

The diarist, John Evelyn, who had previously visited the area in 1654, wrote of the Holy Well that it 'was said to heal many infirmities, as King's Evil [a form of tuberculosis], leprosy, sore eyes etc.' Around ten years later the first volume of the *Transactions of the Royal Society* refers to the Holy Well in respect of 'a long and old fame for healing eyes'. Wall's own prescription, of drinking freely of the waters for days or weeks followed by external use by washing and sometimes periodically dampened cold compresses, was used to treat a whole range of diseases including cancers, leprosy, ringworm, scorbutic rashes (resulting from scurvy), King's Evil and mouth, face and leg ulcers. Another visitor, of 1757, one Benjamin Stillingfleet, remarked, 'I have been at Malvern about ten days where with difficulty I have got a lodging the place is so very full. Nor do I wonder at it, there being some instances of very extraordinary cures in cases looked on as desperate even by Dr Wall.'

One hypothesis suggests that, as the Holy Well is roughly vertically below the Eye Well on the same hillside, it may be fed by the same source as the higher spring, hence eye conditions being listed among its cures. In support of this idea, there is another ancient well which also had a reputation for curing eye disease: although now long-abandoned, the waters of Moorall's Well emerge from the same east-west fault as the Holy Well and Eye Well.

Today the Eye Well can still be reached via a footpath. Dr Wall's analysis had shown that the waters of the Eye Well and Holy Well were of very similar composition; but the Eye Well now is little more than a muddy pool, having become silted up possibly as a result of the construction of the footpath. In *The Wells and Springs of Worcestershire* (1930) a Mr Wickham states that the well was a much more prolific source until the nineteenth century, when an attempt to divert its waters to the then-rectory of Malvern Wells was botched. Apparently, as a workman struck the ground with his pick axe the water disappeared completely. Wickham quotes a Mr Bennett as saying that 'owing to such experiences owners of springs in Malvern are always very careful how they interfere with springs. It is the brashy [ie brittle, fragmented] nature of Malvern rock which causes such an occurrence'.

The Eye Well did eventually reappear, albeit as the muddy pool described above. As *The Wells and Springs of Worcestershire* puts it: 'The water oozes out of the Archaen [Malvernian] rock, but owing to the path functioning as a dam, the "well" is now simply a little morass.' Bennett, again in *Wells*, notes that '[Malvern's] wells issue as "fault springs" ... of remarkable organic and mineral purity, but the contact with Paleozoic and Mesozoic rocks imparts some degree of hardness.' However, he continues, in the case of Holy Well which issues above the fault line 'the water never comes in contact with any rock other than Malvernian and owing to the practical insolubility of the rock, the water from these springs is of exceptional purity and softness.'

By the 1970s the Eye Well was in complete disrepair but, in 1992, efforts were made to clear as much of the surrounding debris as possible, remove the silt and unblock the overflow pipe which passes under the footpath. The well is now looked after by its own Malvern Spa Association Well Warden and is 'dressed' annually for the Well Dressing Festival on May Day bank holiday weekend. It is sometimes to be found decorated with flowers and votive offerings at other times too, such as the equinoxes, Midsummer's Day and other ancient festivals.

The bottling of Holy Well water was first recorded in 1622 and, with two short breaks, continued until 2010 which made it - at that time - the oldest bottled water in the country (beginning around 40 years later than the oldest in Europe, that of Spa). Now, to partake of that pure liquid and ophthalmic wonder-cure, one has to climb up the Malvern hillside with a container to catch the water that still flows from a pipe set into the rock. But at least it's free.

HEAVENLY SIGHT

The stories behind some of the saints connected with vision and eye health

Assistance for the visually impaired has come a long way since the technique of 'couching' cataracts was described three thousand years ago in Babylonia and the treatment of ocular ailments by dubious concoctions of ground-up animal, vegetable and mineral ingredients in Greek and Roman times. We have sophisticated surgical, laser and pharmacological treatments for eye disease, and a huge range of appliances for the improvement of vision. But what to do if all else fails – say a prayer? Perhaps. But to whom?

In Christian tradition there are many saints who are patron of, or connected with, a wide range of medical conditions and medical practitioners. Those saints most associated with sight are St Lucy, St Odilia and, to a degree, St Jerome.

St Lucy, whose name means 'light' is a patron saint of the blind and those with eye problems. Known as the Protector of Eyesight, she is also the patron saint of opticians and ophthalmologists (and firemen, sailors and her home town). Lucy was born in Syracuse, Sicily, in 284AD, of a wealthy Greek Christian family, and martyred during the reign of Diocletian in 304AD. There are several legends about her life, but the main one revolves around her secret vow at a young age to dedicate her virginity to God. Her father had died while she was an infant and her mother, suffering from an incurable disease and ignorant of the vow her daughter had taken, arranged for Lucy to be married, aged fourteen, to a pagan nobleman in order to secure her future.

Lucy persuaded her mother to make a pilgrimage with her to the tomb of St Agatha in Catania to pray for relief of her illness. Miraculously the illness was cured, whereupon Lucy revealed her vow and gained her mother's consent to spend her dowry on the relief of the poor and commit her life to the service of God. As can be imagined, this did not go down well with her betrothed. With Diocletian's persecutions of Christians being in full swing at that time, Lucy's husband-to-be denounced her as a Christian to the local Governor. Initially he ordered Lucy to bow down to idols; which she refused to do. As a punishment the Governor ordered her to be forced into prostitution at a brothel; but as soldiers attempted to carry her away she miraculously became immovable. Sorcerers could not break the spell and, when an attempt was made to burn her was unsuccessful despite covering her in flammable oils, finally she was stabbed through the neck with a sword. Even then a miracle occurred that she lived long enough to receive Holy Communion before she expired, aged just twenty.

Lucy's connection with sight, or the lack of it, comes from another part of the legend, of which there are two versions. In both cases her eyes were considered to be especially beautiful. One story has it that her eyes were put out as part of her torture. The other recounts that her eyes were greatly admired by her suitor and Lucy, fearful that this admiration was sinful, plucked out her eyes herself with a knife and sent them to him on a dish with the message, 'Here hast thou what thou so much desired.' In both variants God, in acknowledgement of Lucy's bravery and religious devotion, restored her eyes, even more beautiful than before. Since the Middle Ages she has usually been depicted in paintings holding a dish containing two eyes.

Lucy's feast day is December 13, one of the shortest – and, therefore, darkest – days in the old Julian calendar. In Sweden this day marks the start of the Christmas celebration, when the eldest daughter of a family dresses in a white robe and wears a wreath-crown of evergreens studded with candles in imitation of the similar crown Lucy used to wear when taking food and alms to fellow Christians hiding from persecution in underground catacombs.

St Odilia (or Odile) of Alsace and Strasbourg was born around 660AD to Adalric, first Duke of Alsace. He and his wife had been trying for some years to have a child; when their prayers were answered Adalric was full of rage that not only was it a girl, rather than the hoped-for boy, but the child was blind. He ordered the baby to be cast out or killed, but the family's nurse spirited her away and looked after her, later to be brought up in a convent. At the age of twelve she was taken to be baptised by St Erhard of Regensburg, Bishop of Bavaria. At the moment of baptism Odilia miraculously became sighted and looked directly at the bishop, who said, 'So, my child, may you look at me in the kingdom of heaven.'

Meanwhile Adalric did have other children. His eldest son, Hugh, learned about his lost sister, found her and brought her home without permission from his father; who, on learning of his son's actions killed him in a fit of anger. As penance he accepted his daughter only to see her flee from home when betrothed by her father to a German duke. Some accounts have Adalric following her but being frustrated in his search by a cave opening up in a mountain to hide her; others have him finding her carrying grain to make food for the poor and giving alms for his soul's sake, upon which he gave her a castle with all its lands and revenues to enable her to found a nunnery. The Hohenburg Abbey was founded at the top of a steep hill, where, in a chapel near the convent church, Odilia was buried after her death on December 13, 720. Her tomb has since been destroyed, but there is still present a shrine which is visited by those afflicted with eye disease. She shares her feast day with St Lucy, and is depicted in paintings mostly as an abbess with a book on which there are two eyes. She is sometimes associated with larkspur, which itself is noted for its herbal healing of eye conditions.

In the early part of the twentieth century a 'Guild of Opticians' was established under the banner of the Guild of St Odilia. The December 23, 1938 edition of the *Catholic Herald* reports briefly on the sixteenth bulletin of the Guild of Opticians, noting, 'Correspondence in the Optician on sterilisation, and a mission dedicated to St Odilia writing and asking for help, are among the items of general interest in this number.'

St Jerome (331 – 420AD), translator of the bible from Hebrew into Latin (the 'Vulgate' bible), is the patron saint of spectacle makers. His patronage arises from his scholarship and the way artists indicated this aspect by depicting him with spectacles – somewhat anachronistically, since he lived around eight centuries before their invention!

There are other saints, major and minor, who traditionally have some association with eyes. The archangel Raphael is a key figure in the Book of Tobit, through whose agency Tobit's blindness is cured. Hence Raphael is considered a patron saint of eye disease and blind people. Saint Hervé (Harvey) of Brittany was born blind and raised in poverty by his poet mother, eventually becoming a monk and later abbot. Hervé, St Alice of Schaerbeek and St Clare of Assisi (founder of the 'Poor Clares') are also invoked regarding eye disease and the blind.

OPTICAL WIT AND WISDOM

Was Dorothy Parker right about girls in glasses; and other optical trivia

Everyone, in all branches of the optical fraternity, will have built up their own stock of tips, tricks and anecdotes during their working life. No doubt there is an optometrist who still recounts how, some years ago, as a pre-registration student, she was being shown the correct way to adjust a frame side by a newly-qualified colleague. His advice of, 'Now don't put too much pressure on...' was followed by a distinct *crack* as the side came away in his hand. The demonstrator, who also learned an important lesson, is the author; who now presents a miscellany of optical advice and anecdotes.

Cataract extraction is the world's most commonly performed operation. But records exist of cataract operations dating back 3000 years, to Babylonian times. The surgeon could charge 10 silver shekels for operating on a freeman or two silver shekels for a slave, the fees being fixed by law. There was quite an incentive for him to hone his skills, as there were also punishments, fixed by law, for unsatisfactory outcomes. If a slave was blinded, the surgeon had to provide a replacement out of his own pocket; but if a freeman lost the sight of an eye, the surgeon's hand was cut off. The GMC advises that it has no plans to introduce similar sanctions.

On the subject of eye disease, Sir Hans Sloane, physician and founder of the British Museum, provided a remedy for a range of conditions which could be a useful over-the-counter product if only someone would market it. Published in *The Lady's Companion* in 1751, 'Sir Hans Sloane's Receipt for Soreness, Weakness and several other Distempers of the Eyes' runs as follows:

'Take of prepared Tutty [an impure zinc oxide] one Ounce; of Lapis Haematites [ferric oxide] prepared, two Scruples; one of the best Aloes prepared, twelve Grains; of prepared Pearl, four Grains. Put them into a Porphyry or Marble Mortar, and rub them with a Pestle of the same stone very carefully, with a sufficient quantity of Viper's Grease, or Fat, to make a Liniment; to be used daily, Morning or Evening, or both according to the convenience of the patient.'

Additionally, to deal with ocular inflammations, 'The Doctor prescribes Bleeding and Blistering in the Neck, and behind the Ears, in order to draw off the Humours from the Eyes; and afterwards, according to the Degree of Inflammation, or Acrimony of the Juices, to make a Drain by Issues between the Shoulders, or perpetual Blister. And for washing the Eyes, recommends cold Spring-Water.' If the inflammation returns, 'The Doctor says drawing about six Ounces of Blood from the Temples by Leeches, or Cupping on the Shoulders, is very proper.'

For spectacle wearers, there is good advice from the *Cyclopaedia of Practical Medicine* (1833). The first paragraph still has particular resonance with today's optical world: 'The absolute necessity of purchasing glasses under the direction of a qualified person, and of not going into a shop at random and taking just what the shopman gives you, has already been pointed out.

'The frames should be of metal ... Steel is probably the best, although some people prefer gold ... The nose piece, or saddle, should be carefully adjusted to fit the nose. 'The lenses themselves may be made of crystal - that is, Brazilian quartz – or crown glass. The crystal is harder than glass, and is therefore less likely to scratch, and is not so liable to get broken. Moreover, it takes a higher polish, and being more refractive, it may be of less thickness than glass. The great difficulty is to get a piece of crystal free from specks and impurities. Dishonest dealers often supply crown glass for crystal. The best way to distinguish between them is to apply a file to the edge of the material; glass cuts readily, but crystal is much harder. Crown glass is very good, and may be used when the spectacles have to be changed often, or when expense is an object.'

One, of course, must have more than one pair of spectacles, and advertisers have long known how to use fashion to promote sales. Aldous Huxley, in his 1923 comic novel *Antic Hay*, satirises the effect of American advertisers on the cultural elite:

'For sport or relaxation, they tell you, as though it was a social axiom, you must wear spectacles of pure tortoiseshell. For business, tortoiseshell rims and nickel earpieces lend incisive poise. For semi-evening dress, shell rims with gold earpieces and gold nose-bridge. And for full dress, gold-mounted rimless pince-nez are refinement itself and absolutely correct.

'Thus we have a law created, according to which every self-respecting myope or astigmat must have four distinct pairs of glasses. Think if he should wear the all-shell sports model with full dress!

'Revolting solecism! The people who read advertisements like that begin to feel uncomfortable; they have only one pair of glasses, they are afraid of being laughed at, thought low-class and ignorant and suburban. And since there are few who would not rather be taken in adultery than provincialism, they rush out to buy four new pairs of spectacles.'

Frame manufacturers considering their next advertising campaign may wish to take note of that passage. The corrective comment to the above, which they probably wish had never been written, is the famous aphorism coined by Dorothy Parker: 'Men seldom make passes at girls who wear glasses.' But perhaps it isn't entirely true. Two young women, from Eau Claire, Wisconsin, are on record as sending the following riposte to Parker in 1947: 'The statement is entirely erroneous. We would be willing to wager that a survey made among repairing opticians would show that Monday is a very busy day for straightening girls' glasses.' There is even an old French proverb that suggests that Parker had it back to front. It runs, 'Bonjour lunettes; adieu fillettes.'

There is much debate about the need for foreign health professionals to be able to speak good English. A prerequisite for dealing with patients, surely; but woefully inadequate for those attempting to have dialogue with Clinical Commissioning Groups (CCG) or their successor bodies. A GP, Gordon Barclay, who was medical advisor to one of the old FHSAs (Family Health Service Authority) that CCGs replaced, was good enough to provide a lexicon of Truststpeak phrases and their translations in *Monitor Weekly* of 29 March and 17 May 1995. Here is a selection (translation in italics):

- Need to get a steer on this - *find people who agree with me*
- Informal sounding board - *I will find people who agree with me*
- I think what you are saying... - *here are my ideas*
- What are the ramifications? - *how much will it cost?*
- No time scale - *never*
- This needs a fundamental review - *no*
- I will get back to you - *forget it*

Two final thoughts. Firstly, what is the opposite of an optometrist? Here's one answer, from *worldofmoose.com* (as reported by *Reader's Digest*): 'My optometrist thinks my eyes will probably improve. Unfortunately my pessimetrist thinks they'll get worse.' And lastly, a note of caution for *Masterchef* addicts. There was a recent case of a man who was chopping vegetables when some herbs flew up into his eye. Now he's parsley-sighted.

I hope that you have enjoyed reading this book.

If you wish to recommend it to others, it can be bought online at the publisher's website, www.FeedARead.com (search for 'How Glasses Caught A Killer').

In the same way, it can also be bought via Amazon and other online booksellers. A review on these websites would be much appreciated in order to promote public awareness of it.

You are also welcome to visit, and comment at, the book's Facebook page, www.facebook.com/howglassescaughtakiller

Thank you!

David Baker

9 781786 103246